'STEMS AND TELECOMMUNICATIONS

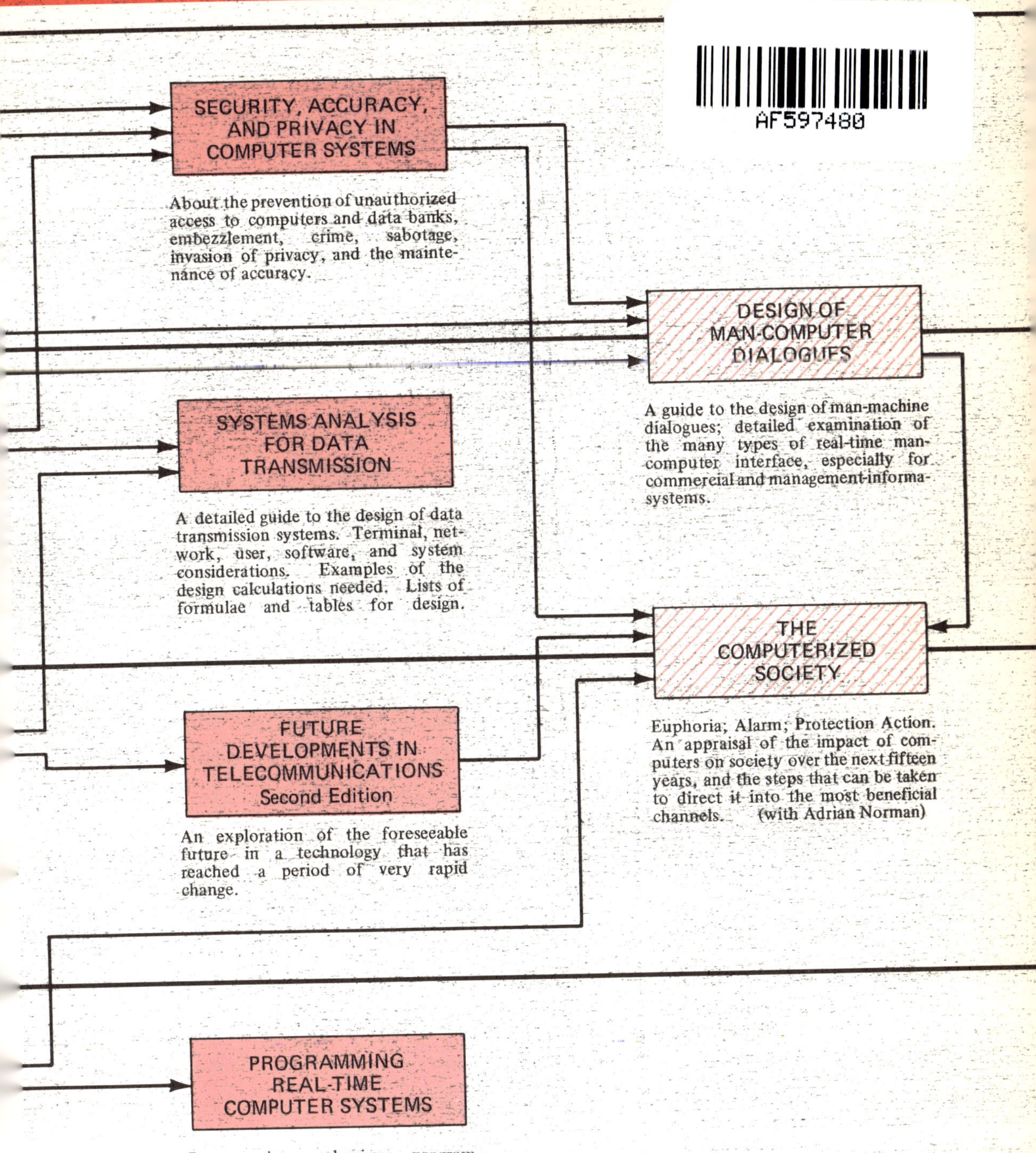

Programming mechanisms, program testing tools and techniques, problems encountered, implementation considerations, project management.

INTRODUCTION TO TELEPROCESSING

Prentice-Hall Series in Automatic Computation

George Forsythe, editor

ANDREE, *Computer Programming: Techniques, Analysis and Mathematics*
ARBIB, *Theories of Abstract Automata*
ANSELONE, *Collectively Compact Operator Approximation Theory and Applications to Integral Equations*
BATES AND DOUGLAS, *Programming Language/One, 2nd ed.*
BLUMENTHAL, *Management Information Systems*
BOBROW AND SCHWARTZ, editors, *Computers and the Policy-Making Community*
BOWLES, editor, *Computers in Humanistic Research*
CESCHINO AND KUNTZMAN, *Numerical Solution of Initial Value Problems*
CRESS, ET AL., *FORTRAN IV with WATFOR and WATFIV*
DANIEL, *The Approximate Minimization of Functionals*
DESMONDE, *A Conversational Graphic Data Processing System*
DESMONDE, *Computers and Their Uses, 2nd ed.*
DESMONDE, *Real-Time Data Processing Systems*
EVANS, ET AL., *Simulation Using Digital Computers*
FIKE, *Computer Evaluation of Mathematical Functions*
FIKE, *PL/1 for Scientific Programmers*
FORSYTHE AND MOLER, *Computer Solution of Linear Algebraic Systems*
GAUTHIER AND PONTO, *Designing Systems Programs*
GEAR, *Numerical Initial Value Problems in Ordinary Differential Equations*
GOLDEN, *FORTRAN IV Programming and Computing*
GOLDEN AND LEICHUS, *IBM/360 Programming and Computing*
GORDON, *System Simulation*
GREENSPAN, *Lectures on the Numerical Solution of Linear, Singular and Nonlinear Differential Equations*
GRUENBERGER, editor, *Computers and Communications*
GRUENBERGER, editor, *Critical Factors in Data Management*
GRUENBERGER, editor, *Expanding Use of Computers in the 70's*
GRUENBERGER, editor, *Fourth Generation Computers*
HARTMANIS AND STEARNS, *Algebraic Structure Theory of Sequential Machines*
HULL, *Introduction to Computing*
JACOBY, ET AL., *Iterative Methods for Nonlinear Optimization Problems*
JOHNSON, *System Structure in Data Programs and Computers*
KANTER, *The Computer and the Executive*
KIVIAT, ET AL., *The SIMSCRIPT II Programming Language*
LORIN, *Parallelism in Hardware and Software: Real and Apparent Concurrency*
LOUDEN, *Programming the IBM 1130 and 1800*

MARTIN, *Design of Man-Computer Dialogues*
MARTIN, *Design of Real-Time Computer Systems*
MARTIN, *Future Developments in Telecommunication*
MARTIN, *Introduction to Teleprocessing*
MARTIN, *Programming Real-Time Computing Systems*
MARTIN, *Systems Analysis for Data Transmission*
MARTIN, *Telecommunications and the Computer*
MARTIN, *Teleprocessing Network Organization*
MARTIN, *Security, Accuracy, and Privacy in Computer Systems*
MARTIN AND NORMAN, *The Computerized Society*
MATHISON AND WALKER, *Computer and Telecommunications: Issues in Public Policy*
MCKEEMAN, ET AL., *A Compiler Generator*
MINSKY, *Computation: Finite and Infinite Machines*
MOORE, *Interval Analysis*
PLANE AND MCMILLAN, *Discrete Optimization: Integer Programming and Network Analysis for Management Decisions*
PRITSKER AND KIVIAT, *Simulation with GASP II: A FORTRAN-Based Simulation Language*
PYLYSHYN, editor, *Perspectives on the Computer Revolution*
RUSTIN, editor, *Debugging Techniques in Large Systems*
SACKMAN AND CITRENBAUM, editors, *On-Line Planning: Towards Creative Problem-Solving*
SALTON, *The SMART Retrieval System: Experiments in Automatic Document Processing*
SAMMET, *Programming Languages: History and Fundamentals*
SCHULTZ, *Digital Processing: A System Orientation*
SCHWARZ ET AL., *Numerical Analysis of Symmetric Matrices*
SHERMAN, *Techniques in Computer Programming*
SIMON AND SIKLOSSY, editors, *Information on Processing Systems: Experiments with Representation and Meaning*
SNYDER, *Chebyshev Methods in Numerical Approximation*
STERLING AND POLLACK, *Introduction to Statistical Data Processing*
STOUTEMYER, *PL/1 Programming for Engineering and Science*
STROUD, *Approximate Calculation of Multiple Integrals*
STROUD AND SECREST, *Gaussian Quadrature Formulas*
TAVISS, editor, *The Computer Impact*
TRAUB, *Iterative Methods for the Solution of Polynomial Equations*
VAN TASSEL, *Computer Security Management*
VARGA, *Matrix Iterative Analysis*
VAZSONYI, *Problem Solving by Digital Computers with PL/1 Programming*
WAITE, *Implementing Software for Non-Numeric Application*
WILKINSON, *Rounding Errors in Algebraic Processes*
ZIEGLER, *Time-Sharing Data Processing Systems*

INTRODUCTION TO TELEPROCESSING

JAMES MARTIN

IBM Systems Research Institute

Prentice-Hall, Inc.   Englewood Cliffs, New Jersey

Current printing (last digit):
17 16 15 14 13

ISBN: 0-13-499814-6

Library of Congress Catalog Card Number: 74-38242
Printed in the United States of America

PRENTICE-HALL INTERNATIONAL, INC., *London*
PRENTICE-HALL OF AUSTRALIA, PTY., LTD., *Sydney*
PRENTICE-HALL OF CANADA, LTD., *Toronto*
PRENTICE-HALL OF INDIA PRIVATE LIMITED, *New Delhi*
PRENTICE-HALL OF JAPAN, INC., *Tokyo*

TO CHARITY

CONTENTS

Appendices:

STATEMENT OF INTENT

It is intended that this book should provide the easiest possible way to learn about data transmission from the beginning. It can be read by nontechnical persons and it is hoped that the book can be enjoyable reading rather than a chore. Many readers will be in the computer industry and the Appendix with its checklists is intended for them.

All of the subject matter in this book is taken from the author's other books but has here been rewritten to form an introduction to the subject. At many points the reader of serious intent will require more detailed knowledge. To provide this, references are given throughout to that chapter in the author's other books where further reading can best continue.

The Appendices give comprehensive summaries and checklists for use in the design and selection of teleprocessing equipment. These may be used by management to assist them in asking the right questions of their data processing staff and systems analysts. For discussion of many of the items on these checklists, the Appendices again refer the reader to a chapter in another book containing details. It is intended that these Appendices will be of value for repeated reference.

James Martin

INDEX OF BASIC CONCEPTS

The basic concepts, principles, and terms that are explained in this book are listed here along with the page on which an introductory explanation or definition of them is given. There is a complete index at the end of the book.

INTRODUCTION TO TELEPROCESSING

1 COMPUTERS OF THE WORLD, UNITE!

In 1980 the computers of the United States alone will transmit or receive 250 billion data transactions over telecommunication lines.

This is a conclusion of a $1 million study carried out for the Datran Corporation by Booz, Allen, and Hamilton. It seems a conservative estimate, for many likely uses of data transmission have not been included and many more than their estimated 2½ million terminals may exist. Today's telephone plant would be inexorably jammed by such a load; thus it is clear that there must be massive construction of new communication facilities in the years ahead. The majority of today's computer personnel will become involved with data transmission, as will the majority of telecommunication engineers. And numerous new personnel must enter this field.

A vast network of telecommunication links spans the industrialized countries of the world, carrying telephone, telegraph, and television signals, news photographs, and radio programs. This network is constantly being expanded and new inventions are increasing its capacity at a breathtaking rate. As yet computer systems have made only minor use of this immense network. For a decade or so, data have been transmitted over communication lines, often in a pioneering, experimental manner. There were many problems in the beginning, but slowly solutions have been found. We have learned to use the links more efficiently and the costs have started to drop. It is now practical for one computer to dial up another computer—just as we dial one another—and transmit information. Even with the speed restrictions of a conventional telephone line, the computer can often code its information so that it sends hundreds or thousands of times as much in a given time as human speakers.

Perhaps more important, *we* shall be able to dial the computers and communicate with them. In offices, shops, factories, and, probably, individual homes, there will be small machines designed to enable men to communicate with distant computers. We shall be able to ask them questions, to interrogate enormous banks of stored information, to perform calculations and to enter data that the computers will store, process, and act upon. In more advanced applications, we are seeing a new type of thinking in which the creative ability of the human user interacts with the enormous logic power of the machine and in which man has access to its vast store of data. This interaction can produce results that neither man nor machine could achieve alone.

Systems described as on-line and real-time are now installed in many organizations. Under these systems data may be entered directly into the computer system from the environment it works with, and results relayed back. The word real-time implies that the response comes back very quickly, usually within 2 seconds or so if the response is to a man and sometimes in a fraction of a second if it is to a machine. A wide variety of devices have been built for feeding data to, and receiving replies from, a distant computer. Such devices are referred to as terminals.

TERMINALS

An ever-growing array of machines can be attached to communication lines for transmitting data to a computer. These can be devices into which data are entered by human operators or devices that collect data automatically from instruments. Terminals designed for human use may permit a fast two-way "conversation" with the computer or may be a remote equivalent of the computer-room input-output devices. The people in Figure 1.1 are using terminals to communicate with a computer. Paper-tape readers and punched-card readers may provide input over communication lines. Printers may provide the output. Figure 1.2 shows non-real-time terminals.

Most computer peripherals can be taken out of the computer room and attached to a communication line. They can have a typewriter added, or a keyboard or screen display; and then they are called conversational terminals.

With a control console, they can be called an off-line data-preparation terminal; or with a cluster of manual-input devices, they can be called a data collection system. Tape cartridges or small disks can be added. Logic circuitry and memory can be added in varying quantities. Microprogramming may be used or even a stored-program mini-computer. The endless variations on these terminal possibilities are further proliferated by the fact that the components can be supplied from many hundreds of peripheral equipment manufacturers. The units are often built in a modular fashion so that, as with a hi-fi system, a variety of different devices may be added. The user's choice can be complex.

The information, whether from automatic devices or from manually operated keyboards, may be transmitted immediately to the computer or may be stored in some medium for transmission at a later moment. The preparation of data, in other words, may be *on line* or *off line*. Readings of instruments, for example, may be punched into paper tape, which is later transmitted to the computer. Similarly, data collected from manually operated devices may be punched into paper tape. In this case, the terminal of the computer is an on-line paper-tape reader. The output may also make use of an interim medium, such as paper tape or punched cards, or it may directly control the environment in question. Very often it is necessary to make a printed copy of the computer output for later analysis. In this case, part of the terminal equipment may be a typewriter or printer.

Table 1.1 lists some of the common types of terminal devices. Many machines use combinations of these facilities.

Teleprocessing is the most rapidly growing area in the exploding data-processing industry. The reason is the power and versatility that the inter-linking of computers can bring, plus the potential benefits to the individual

Table 1.1

Document Transmission Terminals	Human-Input Terminals	Answerback Devices and Displays
Paper-tape readers/punches	Typewriterlike keyboard	Typewriter
Card readers/punches	Special keyboard	Printer
Magnetic card readers	Matrix keyboard	Teleprinter
Badge readers	Lever set	Passbook printer
	Rotary switches	Display screen
Optical document readers	Push buttons	Display tubes
Magnetic-ink character readers	Teleprinter	Light panel
Mark-sensing devices	Telephone dial	Microfilm or film-strip projector
Microfilm	Touch-Tone telephone keyboard	Big display boards
		Graph plotter
Plate readers	Light pen with display tube	Strip recorder
Facsimile machines		Dials
	Coupled stylus	Telephone voice answerback
Magnetic-tape units	Facsimile machine	Facsimile machine
Tape cassettes	Plate reader	
Magnetic disks	Badge reader	

Children in a Brooklyn school "talking" to an RCA computer in New York. The same computer is handling 200 such terminals in 15 widely separated schools. (Photograph from RCA.)

Real-time displays being used during an election broadcast. All of the displays seen are connected to one telephone line.

Figure 1.1. **Operators at Real-Time Terminals.**

Airline agents taking calls from the public about flight bookings. They answer enquiries, find the best flights, and make and change reservations. All the terminals shown are connected via a concentrator to one telephone line. (Photograph courtesy American Airlines.)

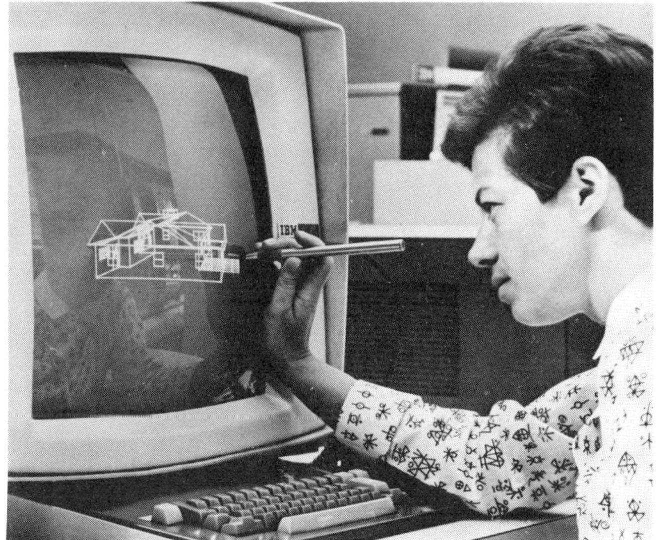

A house being designed with a light pen at an IBM graphics terminal. Teleprocessing is only rarely used today with pictorial graphics because a transmission rate higher than that of a voice line is needed to transmit the images fast enough. This will change with better image encoding.

Figure 1.1. **Continued**

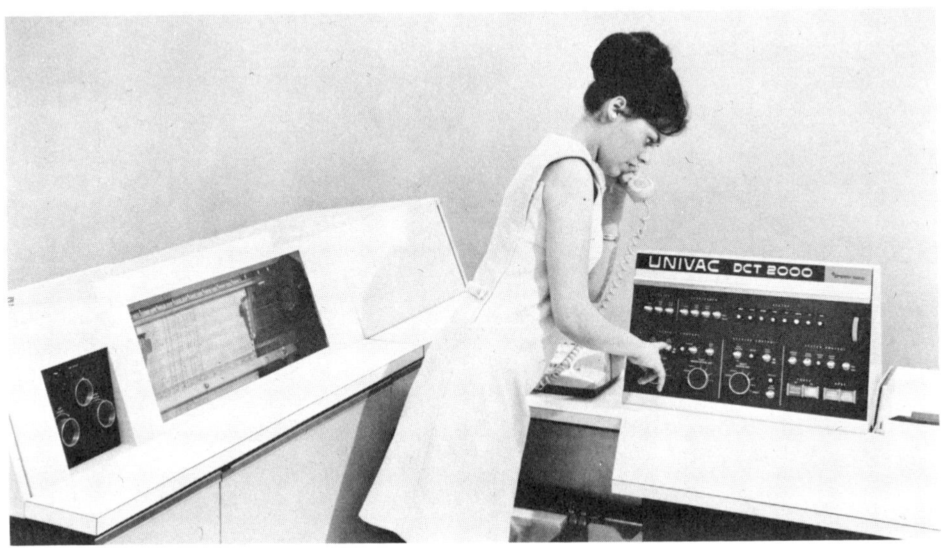

The Univac DCT 200 Data Communication Terminal consists of a 250-line-per-minute card reader, a 200-card-per-minute card reader, a 75-card-per-minute punch and their control unit. These transmit or receive synchronously over a voice line.

An A.T. & T Dataspeed terminal for transmitting and receiving data on paper tape.

Figure 1.2. **Non-real-time terminals.**

An IBM 1255/2770 system for transmitting data from bank checks and other MICR (Magnetic Ink Character Record) documents.

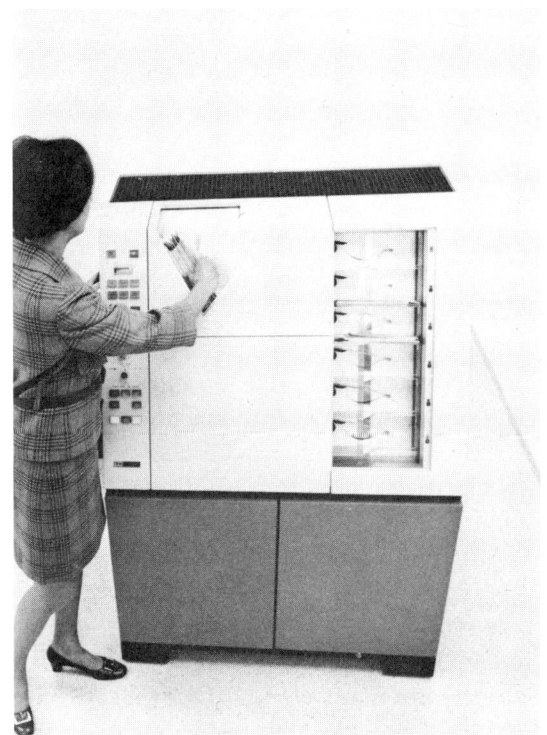

The IBM 7702 Magnetic Tape Transmission Terminal (below) transmits the contents of magnetic tapes to another 7702 or other type of compatible device. Transmission over a voice line is used at speeds up to 300 characters per second.

Figure 1.2 **Continued**

of having this power at his fingertips. In all walks of life and in all areas of industry, the devices connected to distant computers will change the realms of what it is possible for man to do. In centuries hence historians will look back on the coming of computer data transmission as a fundamental step forward in civilization, perhaps eventually having a greater effect on the human condition than even the invention of the printing press.

Data transmission will drop in cost. Long-distance costs will drop much more than short-distance costs, thus having an effect on the organization of nationwide corporations. Intercontinental links will drop much more in cost than national links, and the international corporation will be bound together with ties of satellite communication.

Data transmission will become as indispensable to city-dwelling man as his electricity supply. He will employ it in his home, in his office, in shops, and in his car. He will use it to pay for goods, to teach his children, to obtain information, transportation, stock prices, items from the shops, and sports scores; he will use it to seek protection in the crime-infested streets. The best potentials of data transmission will give man more knowledge, more power, more leisure time; he will have less mandatory travel; his job will be more interesting. In many ways, his life will be richer.

The worst potentials of data transmission conjure up visions of George Orwell's *1984* and will cause us to look harder at issues involving privacy, security and the democratic process.

2 CATEGORIES OF DATA TRANSMISSION SYSTEMS

Data transmission systems are built for a wide variety of purposes and differ accordingly in the way they function.

Probably the most common form of system in the future will be one in which people at terminals communicate with a distant computer. The computer will usually respond to them quickly. Often a dialogue takes place between the terminal user and the remote computer. This type of application did not exist for the first hundred years of data transmission, and therefore many of the traditional ideas are being rethought. Indeed, when a person whose background is data processing approaches data transmission, he does so with a point of view fundamentally different to that of the traditional telecommunications man.

ON-LINE AND OFF-LINE SYSTEMS

Many systems are not real-time in nature but are required merely to move a quantity of data from one point to another. The communication links in this case may be on-line to a computer or off-line. *On-line* means that they go directly into the computer, with the computer controlling the transmission. *Off-line* means that telecommunication data do not go directly into the computer but are written onto magnetic tape or disk, or they are punched into paper tape or cards for later processing.

An on-line system may be defined as one in which the input data enter the computer directly from the point of origination and/or the output data are transmitted directly to where they are used. The intermediate stages of punching data onto cards or paper tape or of writing magnetic tape or off-line printing are avoided.

INTERACTIVE AND NONINTERACTIVE SYSTEMS

Off-line systems are not "interactive." Because no computer is directly connected at the location the data are sent to, no data response will be received from that location, although simple control signals may be received to control the mechanical functioning of the devices and to indicate whether the transmission has been found free of errors.

Some on-line systems are also noninteractive. The computer may merely receive a batch transmission and may have no need to respond to it. Sometimes it may take a hash total at the end of a batch transmission to ensure that no data have been garbled, and so the only interactive response is confirmation of correct receipt of the transmission.

Most transmissions from human operators at terminals are interactive; in fact, it is bad design not to give a response to an operator and to leave him wondering whether his input has reached the computer or not. In inexpensive terminals, however, there may be no mechanism for responding. Such is the case with some factory data-collection terminals into which a worker may insert a machine-readable badge and set some keys or dials. The system may respond with a light or mechanical action indicating simply that it has received the message.

For noninteractive systems or systems that give a very rudimentary response, the data will flow in one direction only. The transmission system is not normally designed to be entirely one way because a small trickle of control signals going in the other direction is needed. Occasionally, in telemetry, the use of radio makes two-way transmission difficult and a purely one-way link is used. An interactive system, on the other hand, can have a high flow of data in both directions. Factors such as this will sometimes affect the data transmission techniques used.

Figure 2.1 lists some common categories of data transmission systems.

QUANTITY OF DATA TRANSMITTED

The quantity of data required to be transmitted varies enormously from one system to another. At one extreme we find whole files being transmitted, and at the other all that is needed is to indicate a YES/NO condition and one bit would suffice. Sometimes a single transaction is transmitted, sometimes a batch of many transactions. When no immediate action is required on data that are gathered, the information may be collected and transmitted in a batch, which is less expensive than sending it a transac-

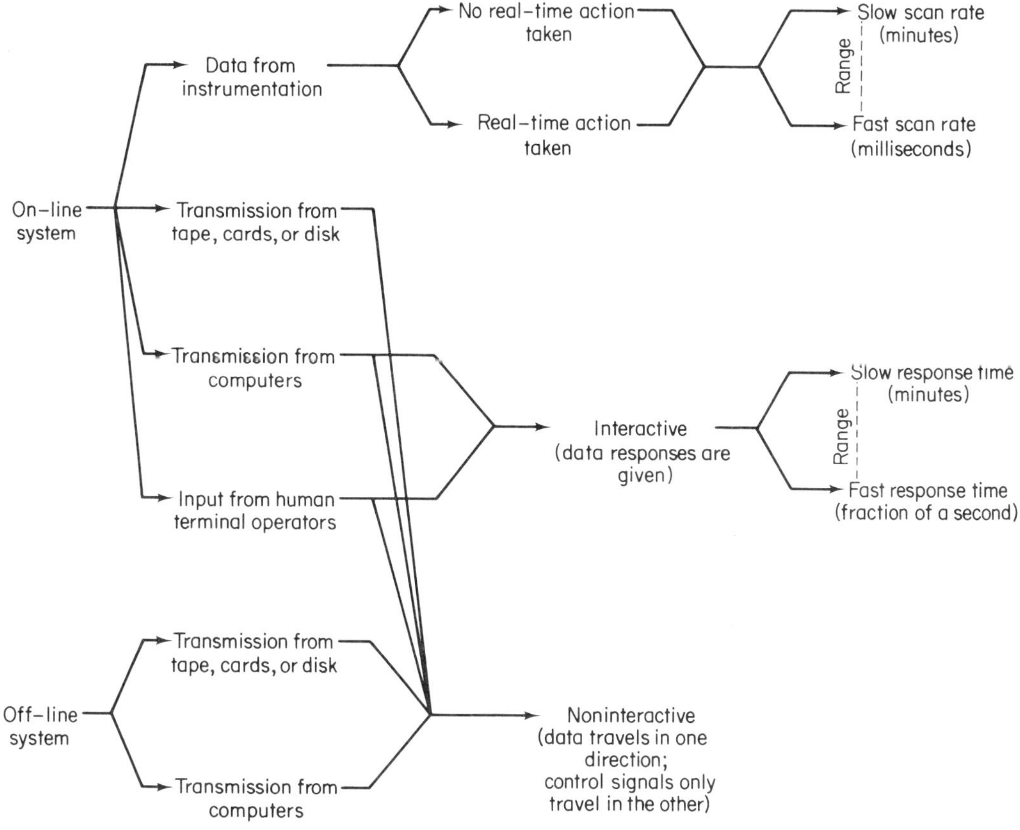

Figure 2.1. Common categories of data transmission systems.

tion at a time. Sometimes the contents of entire magnetic tapes or disks are sent. Payroll data or machine-shop schedules, for example, may be transmitted from one location to another. When a "dialogue" takes place between a terminal user and a computer, the message sizes then depend on the design structure of the dialogue (see Chapter 9). The input to the computer is one statement by the user, and the output is the computer's response, which may range from being a one-character confirmation to a screenfull of data or a printed listing.

TIME FOR TRANSMISSION

Sometimes it is necessary to transmit the data quickly. The speed required depends on the system. A system for relaying one-way messages, like cables, may be required to deliver the message in an hour or so. It would be convenient to have it done faster, but no major economic need to do so exists. When batches of data are sent for batch processing on a distant computer, a delivery time longer than one hour is sometimes acceptable. However, where a man-computer dialogue is taking place, the responses must be returned to the man sufficiently quickly so as not to impede his train of thought. Response times between 1 and 5 seconds are typical. In real-time systems in which a machine or process is being controlled, response times can vary from a few milliseconds to many minutes.

Response time for a terminal operator can be defined as *that time interval from the operator's pressing the last key of the input to the terminal's typing or displaying the first character of the response.* For different types of situations, response time can be defined similarly: *the interval between an event and the system's response to the event.*

Systems differ widely in their response-time requirements, and the response time needed can, in turn, have a major effect on the design of the data transmission networks.

Where systems are not interactive, we might specify a "delivery time" rather than a response time. Delivery time refers to situations in which the data are flowing in one direction and can be defined as *that time interval from the start of transmission at the transmitting terminal to the completion of reception at the receiving terminal.*

The question of how fast the response time must be can have a major effect on the cost of the network [1]. For a dialogue in which the operator must maintain a continuing chain of thought, responses in 2 seconds or less are usually needed. Where the input is a simple inquiry, or a request for a listing, or for a program to be run, the response time may be longer, sometimes much longer. Where machinery is being controlled, a faster response is sometimes necessary; and where one computer is requesting data from another computer, it may be needed very quickly so that the requesting computer itself can achieve a specified response time.

Figure 2.2 shows some of the common requirements for delivery time or response time, and for quantities of data transmitted.

The block labeled "terminal dialogue systems," for example, indicates a response-time requirement of from 1 to 10 seconds, and a message size ranging from one character (usually 7 or 8 bits) to about 4000 characters (around 30,000 bits). A few exceptional cases will extend beyond the block shown (as with all of the blocks on the diagram).

The block labeled "terminal inquiry systems" extends on the time scale to 100 seconds. A wait of this duration would be quite unacceptable in a dialogue situation, but where a storekeeper or foreman (for example) is mak-

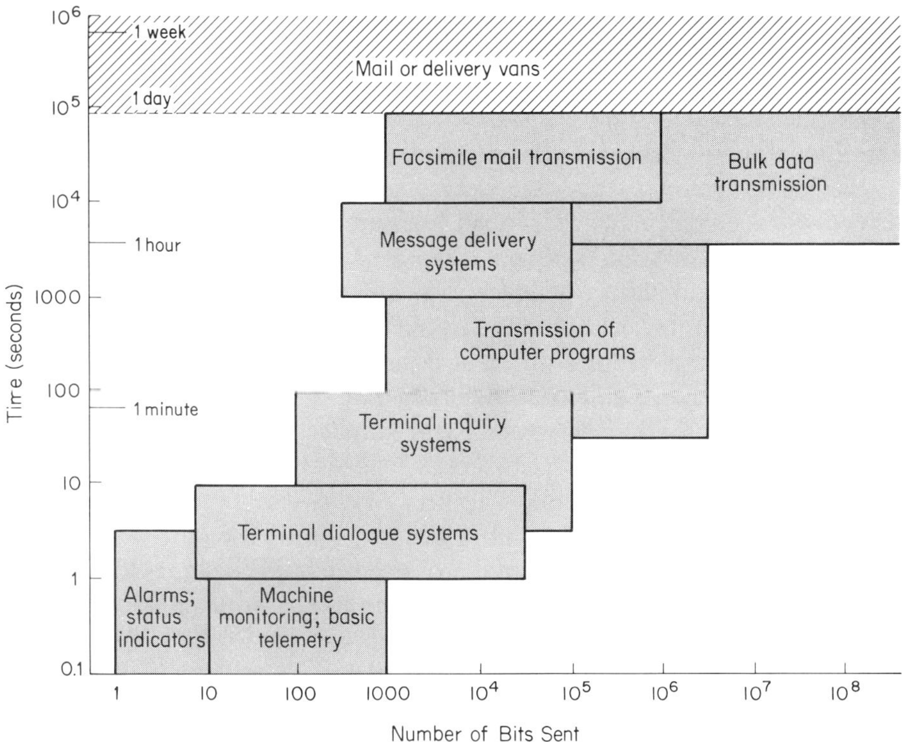

Figure 2.2. Desirable delivery times and typical quantities of data for common uses of data transmission.

ing a simple request for a listing, a wait of a minute may not cause much inconvenience. In all such cases, the cost of providing different response times must be related to the requirements of the user.

Systems for delivering messages or cables, commonly called *message switching systems*, are referred to with the block labeled "message delivery systems." With the apparent deterioriation of the mail services, it seems likely that facsimile mail transmission will grow in popularity. Certainly it will be the fastest way of delivering mail, but delivery times up to one day will be acceptable. The same delivery times will often be acceptable for the bulk transmission of data. Bulk transmission of data extends to the extreme right of the diagram because sometimes very large quantities of data are sent.

There is not much point in using data transmission over a line so slow that the delivery time is more than 24 hours. In these cases, mail, delivery van, or air transportation is usually cheaper.

In the years ahead it is probable that data transmission will be used in areas outside the blocks shown in Figure 2.2. To some extent the areas in which it has been used to date have been determined by the speeds of available transmission lines. The transmission speed equals the delivery time of one-way messages on Figure 2.2 (vertical axis), divided by the number of bits transmitted (horizontal axis). For most of the applications shown in the figure, the speed of teletype and telephone lines—speeds up to, say, 4800 bits per second—is sufficient. Higher-speed lines are required today mainly when a very large bulk of data must be sent.

The computer industry changes fast, and it seems probable that those areas in which a computer responds to a machine or another computer, rather than to a person, will require different transmission facilities. One computer may request from another distant one a program or data for immediate processing. One computer will pass jobs on to another because it does not have the capability to handle them itself, or for load-sharing reasons. The requirement for access time in obtaining data from disks—the time taken to "seek" and read a record—may in the future become the time requirement for obtaining data from a distant computer over telecommunication links. Small local computers will sometimes pass their compiling work on to bigger distant machines with better facilities. The initial program loading of small local computers will sometimes be done by a large remote computer.

The computers and file units that are connected by cabling in the machine room today may be connected by telecommunication links tomorrow if channels of suitable speed are available.

TIME-SHARING SYSTEMS

Most systems with manually operated terminals are time-shared, meaning that more than one user is using them at the same time. When the machine pauses in the processing of one user's item, it switches its attention to another user.

The term "time sharing," however, is commonly used to refer to a system in which the users are *independent*, each using the terminal as though it were the console of a computer and entering, testing, and executing programs of their own at the terminal. Each user feels as though he were the only person using the system. The programs of one terminal user are quite unrelated to those of other users.

In many systems the users do not *program* the system at the terminal, and neither are the users entirely independent. They are each *using* the programs in the computer in a related manner. They may, for example, be insurance or railroad clerks possibly using the same programs and possibly the same file areas. In many time-shared systems, however, the computing facility is being divided between separate users who can program *whatever they wish* on it, independently of one another. Four categories of on-line systems, differing by the degree of independence of the users, are in common use.

1. Systems that carry out a carefully specified and limited function—for example banking systems, in which all terminal users can update the same files.
2. Systems for a specific limited function in which the user has personal independent files, or in which shared files can be read but not updated by general terminal users.
3. Systems in which programmers can program anything they wish at the terminals, providing they all use the same language, an interpreter or compiler for this being in the machine. Normally each user has the same type of terminal.
4. Systems in which programmers can program anything by using a variety of different languages. Often a wide variety of terminals is permitted also.

The last two types of system are, in effect, dividing up the computing facility "timewise" and giving the pieces to different programmers to do what they want with the pieces. The files are also divided according to use. It cannot be known beforehand how much space in the files the various users will occupy. This is quite different from category 1, in which the file size and organization are planned in detail.

TIME-SHARING COMMUNICATION LINES

The reason why time sharing is so important in computers is that the human keying rate and reading rate are very much slower than computer speeds. Furthermore, the human being requires lengthy pauses to think between transactions. The key to using the computer efficiently for real-time dialogue operations is to make it divide its time between many users. The same is true with telecommunications. A voice line could transmit many more bits than a terminal operator generally employs in an alphanumeric dialogue with a computer. The key to using it efficiently is to time-share the line between many users. There are several ways of doing so, as we will discuss in later chapters.

The need to share communication lines will become greater in the future, as facilities capable of transmitting higher bit rates will be employed. Today's telephone lines are commonly made to carry data rates of 2400 or 4800 bits per second. However, on the Bell System T1 carrier, of which millions of miles are already installed, the telephone voice channel is equivalent to 56,000 bits per second.[1]

The degree of independence of users on a shared communication line varies. On some shared lines, all users must employ the same terminal type with the same line-control procedure. On others, they can be different but must use the same character code. With some systems they can be entirely independent, using different codes. The users may all be of one type, in one organization—reservation agents linked to an airline system for example. On the other hand, they may be different types of users but may share communication facilities installed within one organization. Finally, they might be entirely independent users sharing public data transmission facilities. It is in the latter case that the largest savings may result, and eventually public data networks will play a very important part in a nation's data processing.

TYPES OF SYSTEM

Table 2.1 indicates some of the types of systems that use data transmission, saying whether they are likely to be on line or off line, in plant or out plant, and interactive or not. Some exceptional systems differ from the categorization shown in this table. The two bottom lines may remain the largest categories and have been expanded in Tables 2.2 and 2.3.

[1]See the author's *Telecommunications and the Computer*, Chapter 15, Prentice-Hall, Inc, 1969.

Table 2.1 A CATEGORIZATION OF COMMON DATA TRANSMISSION SYSTEMS. THE TWO BOTTOM LINES ARE EXPANDED ON IN TABLES 2.2 AND 2.3.

	On Line or Off Line	Interactive?	Typical Response Time (seconds)	Typical Delivery Time	In Plant or Out Plant
Data collection system	ON or OFF	YES or NO	1–5 sec		IN
Alarm system	ON	YES or NO	1–20 sec		IN or OUT
Machine monitoring	ON or OFF	YES or NO	Wide variation		IN or OUT
Telemetry	ON or OFF	YES or NO		Short	IN or OUT
Process control	ON	YES or NO	1 sec–10 min		IN
Batch transmission	ON or OFF	NO		Dependent on batch or job size	OUT
Load-sharing batch system	ON	NO		Dependent on batch or job size	OUT
Remote job entry	ON	YES		Dependent on batch or job size	IN or OUT
Intercomputer transmission	ON	YES	0.1–1 sec		OUT
Interactive programming	ON	YES	1–5 sec		IN or OUT
Message switching	ON or OFF	NO		½ to 8 hr	OUT
Commercial real-time systems	ON	YES	1–5 sec		IN or OUT
Information utilities	ON	YES	1–5 sec		OUT

Table 2.2. SOME OF THE MANY TYPES OF COMMERCIAL REAL-TIME SYSTEMS

Airline reserve systems	Information-retrieval systems
Banking systems	Library catalog systems
Sales inquiry systems	Museum catalog systems
Sales-order entry system	Hotel-booking systems
Point-of-sale data collection systems	Stores and inventory control systems
Credit information systems	Production data collection systems
Electronic fund-transfer systems	Management inquiry systems
Stockbroker information systems	Engineering design aids
Hospital information systems	Terminals for statistical services
Text-editing systems	Terminals for financial analysis
Document-retrieval systems	Terminals for specialized professional functions

Table 2.3. A FEW OF THE MANY POSSIBLE CATEGORIES OF INFORMATION UTILITIES*

Type	General Purpose or Subscriber Oriented	Retrieval or Processing	Digital or Graphical	Small, Medium, or Large Subscriber	Local, Regional, or National
1. Savings account processing	S	P	D	S, M, L	L, R
2. Stock brokerage information	S	R	D	S, M, L	N
3. Travel service	S	P	D	S, M, L	N
4. Professional billing	S	P	D	S	L
5. Engineering problem solving	G	P	D, G	S, M, L	L, R, N
6. Graphical design	G	P	G	S, M, L	L, R, N
7. Document retrieval	G	R	G	S, M, L	N
8. Data retrieval	G	R	D	S, M, L	L, R, N
9. General purpose	G	P	D, G	S, M, L	L
10. Credit information	S	R	D	S, M, L	L, R, N
11. Financial exchange	Multiple S	P	D	S, M, L	L, R, N
12. Hospital and medical	S	P	D, G	S, M, L	L, R
13. Educational and teaching	G, S	P	D, G	S, M, L	L, R, N
14. Hotel reservations	S	P	D	S, M, L	N
15. Interairline reservations	S	P	D	S, M, L	N
16. Railroad information	S	P	D	S, M, L	N
17. Publishing	S	P	D, G	S, M, L	L, R, N
18. Radio and TV time brokerage	S	R	D	S, M	N
19. Retail and distribution	S	P	D	S, M	L, R
20. Insurance	S	P	D	S, M	L, R
21. Merchandising and advertising	G	P	D	S, M	N
22. Public survey and polling	G	P	D	S	N
23. Sport and theater tickets	S	P	D	S	L, R, N
24. Tax service	G	P	D	S, M	L
25. Labor negotiations	S	P	D	S, M, L	N
26. Career and employment information	G	P	D	S, M, L	N
27. Legal citation service	S	P	D	S, M, L	N
28. Post Office	S	P	D	L	N
29. Marketing research	G	R	D	M, L	N
30. Criminal intelligence	S	P	D	L	N
31. Typing and editing service	G	P	D	S, M, L	N
32. Personal data services	G	P	D	S	L, R, N
33. Library catalogs	G	R	D	S, M, L	L, R, N
34. Lottery service	S	P	D	M, L	L, R, N
35. Terminals for surveyors	G, S	P, R	D, G	S	N
36. Terminals for engineers	G, S	P, R	D, G	S	N
37. Terminals for electronics designers	G, S	P, R	D, G	S	N
38. Terminals for architects	G, S	P, R	D, G	S	N
39. Terminals for lawyers	G, S	P, R	D	S	N
40. Terminals for statisticians	G, S	P, R	D	S	N
41. Terminals for doctors	G, S	P, R	D	S	N
42. Terminals for other professional groups	G, S	P, R	D, G	S	N

*Reproduced with permission in modified form from *Information Utilities* by Richard E. Sprague, Prentice-Hall, Inc., Englewood Cliffs, N.J., 1969.

3 THE COMMUNICATION CHANNEL AND ITS CAPACITY

Imagine a fireman with a long hose going into a building. Unknown to the chief of the brigade, the fireman is trying to transmit data down his hose. He is not really a fireman but an espionage agent who has an accomplice inside the building that the hose goes into.

He attempts to send the data by means of a piston. As he pushes and pulls the piston at one end of the hose, the pulses are transmitted to a receiving piston at the other end. If the hose were absolutely rigid and the water in it absolutely incompressible, the movement of the receiving piston would follow the movement of the transmitting piston exactly. Again, if the water had no viscosity and moved completely without friction, the piston would be able to transmit at a very high speed. However, the hose is not rigid. It is slightly elastic and the water with air bubbles in the hose is slightly compressible, so the receiving piston does not follow the movement of the transmitting piston exactly. Furthermore, there are viscosity and friction; therefore the piston cannot move and transmit at limitless speed.

A communication line has properties that are loosely analogous to these. Electrical properties called "capitance," "resistance," and "inductance" cause it to distort the data transmitted, just as the compressibility, friction, and viscosity cause the fireman's pulses to be distorted.

A clean square-edged data pulse becomes distorted because of these factors as it travels down the communication line. A pulse train starting off

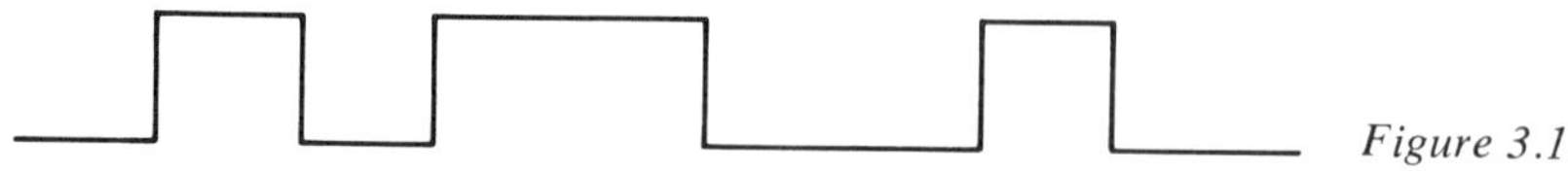

Figure 3.1

as shown in Figure 3.1 ends up as shown in Figure 3.2, with all the voltage transitions slurred.

Figure 3.2

If our James Bond fireman were to attempt to transmit at a fairly slow speed, say one pulse every 5 seconds, the receiving piston would follow the movement of the transmitting piston faithfully enough to recognize each pulse. As he increased his transmission speed, however, the signal distortion would become greater. If he were to transmit at two pulses per second, his accomplice might be able to receive this correctly. At ten pulses per second, sensitive and sophisticated receiving equipment would be needed to detect the pulses without error. Similarly, on a communication line, the natural capacitance and inductance of the line would have a more serious effect, the higher the rate of transmission. (See Figure 3.3.)

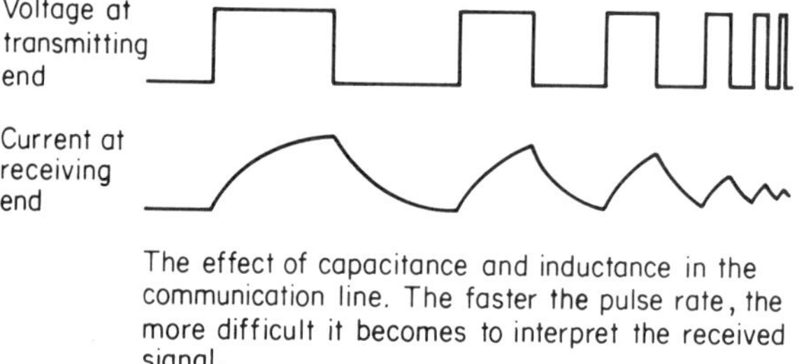

The effect of capacitance and inductance in the communication line. The faster the pulse rate, the more difficult it becomes to interpret the received signal.

Figure 3.3

Furthermore, the impulses fade away as they travel down the line. The fireman might transmit quite successfully at 5 bits per second over a 1000-foot hose, but if the hose were 3 miles long, the same techniques would not work. A very slow bit rate only would be detectable. (See Figure 3.4.)

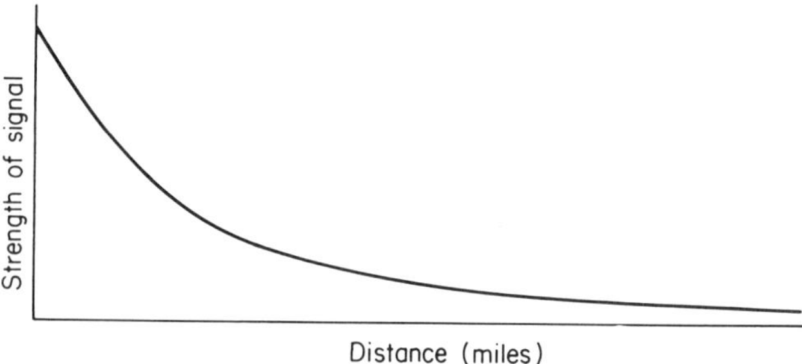

Figure 3.4

THERMAL NOISE

In addition, there is noise on the line. Suppose that the fire hose is vibrating because of the motion of the nearby pump. At high-transmission speeds, the strength of the received pulses becomes comparable in magnitude with this vibration "noise," and errors in the interpretation of the data will occur. In all electronic circuitry, there is a steady continuing background of random noise. It is sometimes called *thermal noise*. The atoms and molecules of all substances vibrate constantly in a minute motion that causes the sensation of heat. The higher the temperature, the greater the vibration. As the atoms vibrate, they send out electromagnetic waves; and because there are many atoms, we have a chaotic jumble of electromagnetic waves of all frequencies. This forms an unavoidable noise background to all electronic processes. We have to send data signals along with this background of a small but ceaseless random variation in signal strength (Figure 3.5).

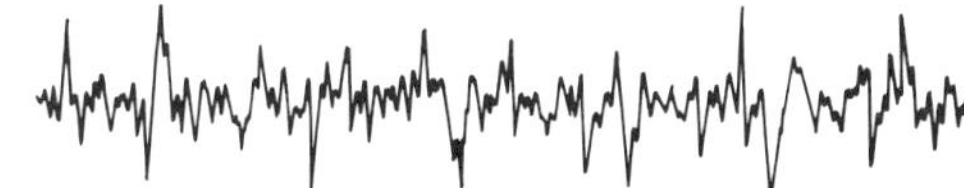

Figure 3.5

When we can hear it, it sounds like a hiss. If the volume of an FM radio is turned up full when no program is being received, one can hear the hiss of this noise.

If the signal falls in strength too much, then it becomes irretrievably mixed up with the thermal noise. Once this happens, the two can never again be separated. If the signal is amplified, the noise will be amplified with it. Some processes in nature are irreversible and this is one of them.

It follows that if we transmit too fast, as at the right-hand side of Figure 3.3, or if we transmit too far, as at the right-hand side of Figure 3.4, then the signal drowns in the noise of Figure 3.5. The longer the transmission distance, the more we are restricted in the speed that can be permitted.

A DIGITAL TRANSMISSION CHANNEL

Given these factors, how would one build a channel for transmitting digital information only? One wants to transmit as much data as possible over, say, an ordinary pair of wires, sometimes for long distances.

Suppose that our fireman did want to transmit data over a fire hose 3 miles long. (Forgive us if we stray somewhat beyond the tenets of Ian Fleming.) Furthermore, he wants to transmit at a high speed. The solution is to build a bit repeater at intervals in the hose. The repeater is a power-driven device that detects the bits being sent and then retransmits them with their original strength and sharpness. The process is shown in Figure 3.6. It suc-

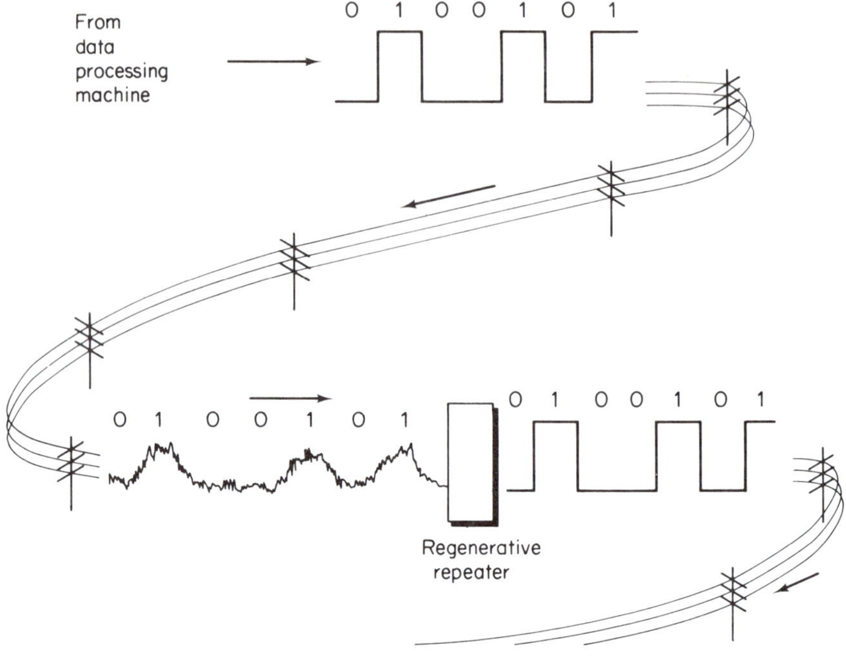

Figure 3.6

ceeds because it detects the bit pattern and recreates it. It catches the bit stream soon enough, before it is submerged in the noise, and then separates it from the noise by creating it afresh.

A very high bit rate could be transmitted provided that the repeaters are sufficiently close together to catch the bits before they are lost in noise. In practice, on a communication line, the repeaters can be small, solid-state devices, not too expensive if mass-produced with the latest technology in sufficient quantity.

We might refer to the line illustrated in Figure 3.6 as a *digital line* because it is designed to transmit bits. It could not transmit music or speech unless these were somehow converted into bits.

TELEPHONE LINES Unfortunately, all telephone lines other than those recently constructed do not work like the line in Figure 3.6. Like firehoses, they were not designed to transmit data bits. Their function was to transmit the human voice. Hence they do not use digital repeaters. When the signal becomes faint, they merely amplify it and so amplify the noise and distortion with it.

In order to transmit the human voice, we have to send a continuous range of frequencies. We hear this range of frequencies coming through the air when somebody speaks.

Light, sound, radio waves, and signals passing along telephone wires are all described in terms of frequencies. In all these means of transmission, the instantaneous amplitude of the signal at a given point oscillates rapidly, just as the displacement of a plucked violin string oscillates. Rate of oscillation is referred to as the frequency and is described in terms of cycles per second.

With light we see different frequencies as different colors. Violet light has a higher frequency than green; green has a higher frequency than red. With sound the higher frequencies are heard as higher pitch. A flute makes sounds of higher frequency than a trombone. Normally the light and sound reaching our senses do not consist of one single frequency but of many frequencies or a continuous band of frequencies all traveling together. A violin note has many harmonics higher than the basic frequency with which the violin string is vibrating. The human voice consists of a jumble of different frequencies. When we see a red light, it is not one frequency but a collection of frequencies that combine to give this particular shade of red. The same is true with the electrical and radio signals of telecommunications. Usually we will not be discussing one single frequency but a collection, or a band, of frequencies occupying a given range.

THE SPEECH SPECTRUM

The human ear can detect sounds over a range of frequencies; in other words, it can hear sounds of different pitch. A sensitive ear can hear sounds of frequencies ranging from about 30 cycles per second up to 20,000 cycles per second, although most people have a range somewhat less than this.

When we refer to a sound of a given frequency, we mean that the air is vibrating with that number of oscillations per second. In order to transmit this sound, the microphone of a telephone converts the sound into an equivalent number of electrical oscillations per second. The telephone channels over which we wish to send data are, then, designed to transmit electrical oscillations of a range equivalent to the frequencies of the human voice, although these frequencies are often changed for transmission purposes.

In fact, the telephone circuits do not transmit the whole range of the human voice. It was found that this was unnecessary for the understanding of the speech and the recognition of the speaker. Figure 3.7 illustrates the characteristics of human speech and shows that its strength is different at different

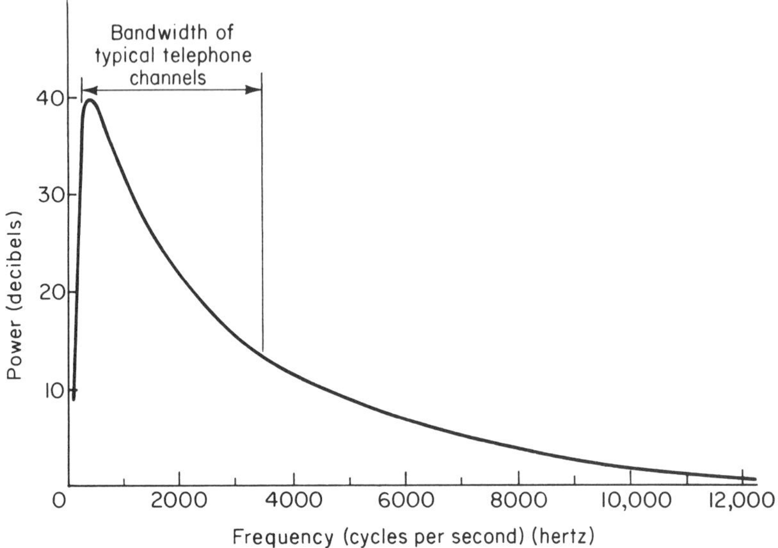

Figure 3.7. The spectrum of human speech. In order to transmit speech so that the speaker is recognizable and understandable only the range indicated need be sent.

frequencies. Most of the energy is concentrated between the frequencies 300 and 3400 cycles per second, and the telephone company, being very concerned about money, transmits only this range of frequencies. This is enough to make the human voice intelligible and the speaker recognizable. When telephone company signals travel over lengthy channels, many may be packed together electronically so that one channel can carry as many as possible. More telephone conversations can be packed together if the upper frequencies shown in Figure 3.7 are sliced off.

ANALOG VERSUS DIGITAL TRANSMISSION

Thus there are two basically different ways in which information of any type can be transmitted over telecommunication media. It can be *analog* or *digital.*

"Analog" means that a continuous range of frequencies is transmitted. Sound and light both consist of such a continuous range. A hi-fi enthusiast strives to reproduce accurately a continuous range from 30 to 15,000 or for people with very good ears 20,000 cycles per second. The frequencies near 20,000 will probably only be heard by the passing bats. If we wanted to transmit high-fidelity music down the telephone wires into your home (which is technically possible), we would send a continuous range of frequencies from 30 to 20,000. The current on the wire would vary *continuously* in the same way as the sound you hear.

"Digital" transmission, on the other hand, means that a stream of ON/OFF pulses are sent, like the way in which data travel in computer circuits. The pulses are referred to as *bits*. It is possible today to transmit at an extremely high bit rate.

Figure 3.8 shows an analog signal and a digital signal. A transmission

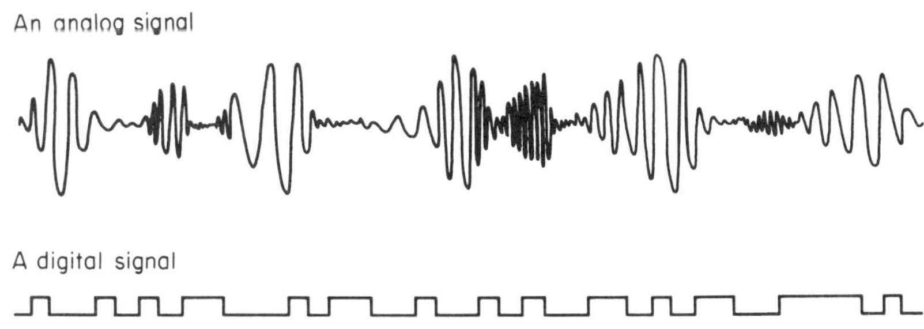

Figure 3.8. Information is transmitted in either an analog or digital form.

path can be designed to carry either one or the other. This applies to all types of transmission paths—wire pairs, high-capacity coaxial cables, microwave radio links, satellites, and the new transmission media, such as waveguides and lasers [1]. If the path is designed to be analog, it will use amplifiers somewhat similar to that in a hi-fi unit for increasing the signal strength. If it is digital, it will use regenerative repeaters, as in Figure 3.6 to reconstruct the bits and pass them on.

MODEMS AND DATA SETS

Unfortunately, many of the communication lines over which we wish to send data are designed for analog transmission, not digital. The telephone channel reaching our home today is an analog channel, capable of transmitting a certain range of frequencies. If we send computer data over it, we have to convert that digital bit stream into an analog signal, using a special device known as a *modem*.

We will discuss modems in Chapter 5. Let us merely note here that a modem is used to convert the square-edged bit stream that leaves the data-processing machine into a range of frequencies suitable for traveling over the analog communication line; then, at the other end of the line, a similar modem converts this range of frequencies back into a bit stream the same as the original. The modem tailors the signal as best it can to fit into the range of frequencies that the communication line handles without undue distortion. The modem is sometimes called a *data set.*

Not only are *telephone* lines analog, but many of the other types of telecommunication links that computers use are as well. Most telegraph lines, and most wideband lines of higher capacity than telephone lines, are analog today. Most of the microwave radio links spanning the country operate in an analog, not digital, fashion. These links, therefore, must also employ modems when they transmit digital signals.

If microwave links or any other communication facility were designed specifically for data transmission, as many will be in the future, then they would be digital in operation, with digital repeaters, and thus would not require modems.

BANDWIDTH

The different physical media used for telecommunications vary widely in their transmission capacity. A coaxial cable can transmit far more information than a simple pair of wires, for example. A digital transmission link is designed for a given data rate, and its capacity will be referred to in terms of numbers of *bits per second.* An analog link, on the other hand, can carry differing data rates, depending on the characteristics of the modem that is used. Its capacity will be referred to in terms of *bandwidth.*

Bandwidth, one of the most important terms in telecommunications, refers to the *range of frequencies that a channel can transmit.* If the lowest frequency a channel can transmit is f_1 and the highest is f_2, then the bandwidth of that channel is $f_2 - f_1$.

Bandwidth is quoted in cycles per second or hertz. "Hertz" means exactly the same as cycles per second. It is a more modern term that replaced cycles per second in the early 60s as the unit of frequency or bandwidth. Technicians everywhere defend their status by inventing new technical words to stall the uninitiated. Kilohertz means thousand cycles per second, and megahertz means million cycles per second.

The bandwidth of a telephone channel is about 3 kilohertz. It can transmit frequencies from about 300 to 3300 hertz. Often the electronics raises the frequency band to high frequencies, say 80,300 to 83,300 hertz, so that it can be transmitted over a particular channel. This does not change the *bandwidth,* which remains 3 kilohertz.

Bandwidth, then, says nothing about the frequency of transmission; it only indicates the range of frequencies.

The capacity of a channel for carrying information is proportional to its bandwidth. A channel of bandwidth 30 kilohertz can carry ten times as many bits per second as the telephone channel of 3 kilohertz.

Imagine a phonograph record with data recorded on it in the dots and dashes of Morse code. If we double the speed of play of the record, we halve the time needed to relay this coded data. Doubling the speed doubles the frequencies of the sound, as well as the bandwidth used. We would say that the sound is higher in pitch. We could play the record at many times its normal speed and relay the data fast, providing that there is some way of interpreting the squeaks that result. When we exceed a certain speed, the sounds will no longer be audible because we have exceeded the frequencies detectable by the human ear. The human ear has a limited bandwidth also.

TOO HIGH A DATA RATE

If we attempt to send bits over a given bandwidth at too high a data rate, they become blurred and uninterpretable. What happens is shown in Figure 3.9.

In this illustration a repetitive bit pattern is being sent: 001000010000-10001. . . . It is sent in the form of square-edged pulses, as shown at the top of the diagram, at a rate of 2000 pulses per second.

To transmit this pulse train exactly, with no distortion at all, would require an infinite bandwidth. Figure 3.9 shows what happens at progressively smaller bandwidths. At a bandwidth of 4 kilohertz, the pulses suffer a little distortion, but not much. At 2500 hertz it is worse. At 1300 hertz the bit pattern is still recoverable. At 900 hertz an appropriate electronic circuit could still recover it, but at 500 hertz it is too distorted to be able to recreate the original with certainty. (The author's *Telecommunications and the Computer* [2] discusses how this diagram was derived.)

Much the same as in Figure 3.9 would happen to the fireman's pulses if he attempted to transmit too fast down his hose of limited bandwidth.

THE MAXIMUM CAPACITY OF A CHANNEL

Transmitting the square-edged pulse train in its original form, as in Figure 3.9, is referred to as *baseband* signaling. On most channels other than short ones, the pulse train is converted into a more complex form by the modem before transmission. We will discuss this process in Chapter 5. No matter how cleverly we manipulate the bits, how-

Bits: 0 1 0 0 0 0 1 0 0
Pulses before transmission
Bit rate: 2000 bits per second
Pulses after transmission
(1) Infinite bandwidth
(2) Bandwidth 4 kiloHertz
(3) Bandwidth 2500 Hertz
(4) Bandwidth 1700 Hertz
(5) Bandwidth 1300 Hertz
(6) Bandwidth 900 Hertz
(7) Bandwidth 500 Hertz

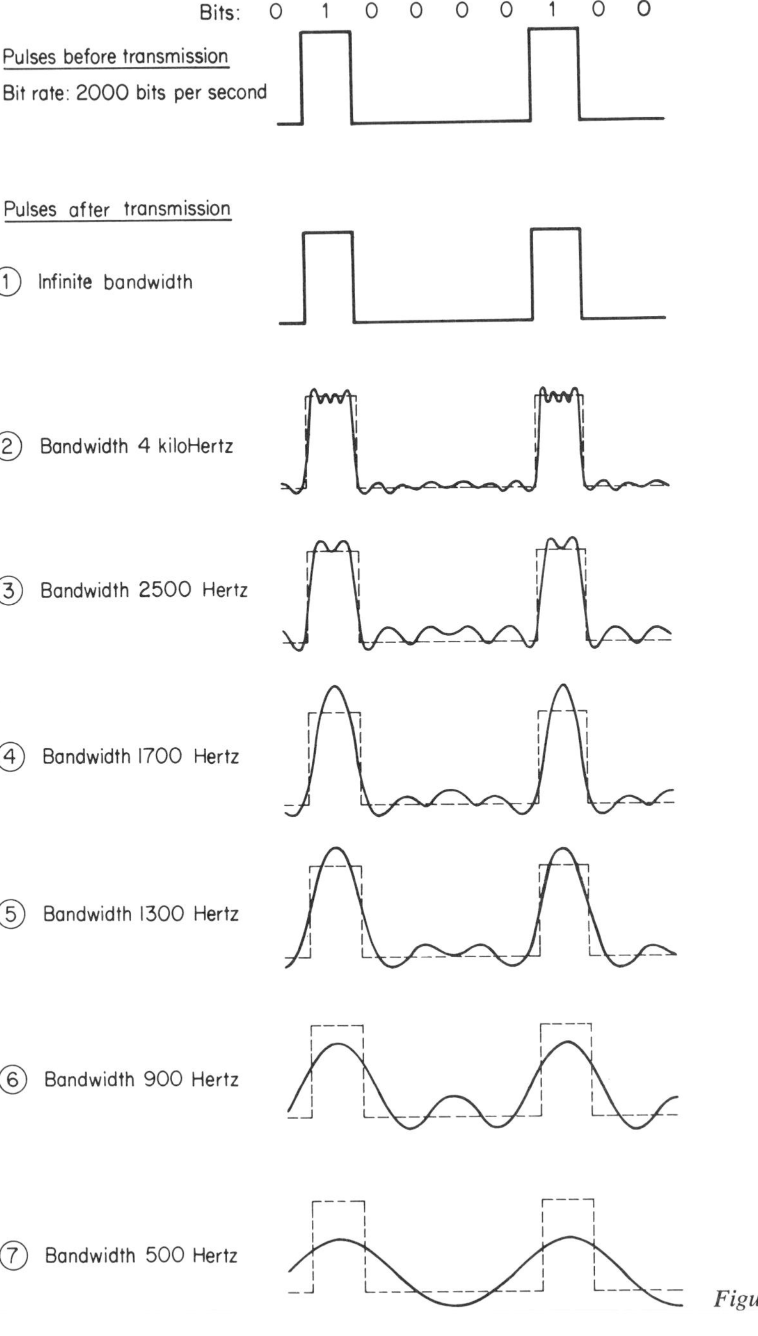

Figure 3.9

ever, we cannot achieve more than a certain maximum bit rate on a given channel.

The ultimate theoretical maximum is given in a theoretical law, as fundamental as the laws of thermodynamics. This is Shannon's law [3]. It states that the maximum capacity of a channel of bandwidth W hertz is:

$$W \log_2 \left(1 + \frac{S}{N}\right)$$

where S = the power of the signal
N = the power of the thermal noise

For an unpredictable bit sequence, Shannon's rigorous mathematical proof shows that there is no possible way of exceeding this quantity of information for these channel parameters. An engineer can design very ingenious modulation techniques and elaborate coding systems, but try as he might he will never send more than this number of bits over the channel unless he increases either the bandwidth available or the signal-to-noise ratio.

It is not unknown for an ingenious inventor to propose a scheme that would do better than the foregoing equation. However, so fundamental is Shannon's law that any such plausible scheme may be treated with the same attitude as inventions for perpetual motion machines. Somewhere in the scheme there is a flaw, and one can say with assurance that it will not work.

In fact, the highest bit rates used in practice on today's lines are often no better than one-third of the theoretical maximum. Feeding figures from a typical telephone line into the preceding equation, we would conclude that it could theoretically carry 25,000 to 30,000 bits per second. In practice, we rarely achieve more than 9600 bits per second.

The reader should note, however, that the achievable data rate in bits per second is substantially higher than the bandwidth in hertz.

ADVANTAGES OF DIGITAL LINES

Most of the world's telecommunication plant is analog in operation at present. Much of it will remain so for years to come because of the billions of dollars tied up in such equipment. However, the technology is rapidly evolving, and it is now clear that there are major advantages in *digital* rather than analog transmission. So great are the advantages that even signals that are analog in nature, such as telephone conversations, music, and Picturephone signals, are beginning to be sent digitally. This portends a major drop in the cost of data transmission [4].

The key factor is that an extremely large bit rate can be achieved if *digital* repeaters such as that in Figure 3.6 are placed sufficiently close on the line. The IBM 2790 system, for example, uses repeaters every 1000 feet on an inexpensive wire pair line for use in factories. By doing so, half a million bits per second are transmitted. The design of this line is conservative; in fact, a rate of several million bits per second could be achieved. The Bell System is similarly using wire pairs with regenerative repeaters for telephone transmission. A wire-pair line system called the *T1 carrier* transmits 1.5 million bits per second, and this is made to carry 24 telephone calls at the same time, each using 56,000 bits per second [5]. Their *T2 carrier* can also use wire pairs, and it transmits 6 million bits per second. This carrier can carry 96 telephone calls or one Picturephone call. Unfortunately, no public tariff is available at the time of writing to permit the data transmission user to employ these bit streams directly.

The direct use of digital lines is the exception rather than the rule today because the common carriers are slow to change their analog technology. Digital transmission will be increasingly common in the future, and by the end of the 1970s most transmission lines other than the local loops from the telephone office to the subscriber will operate digitally.

4 CATEGORIES OF COMMUNICATION LINES

This chapter discusses the types of communication lines that are available. Tables at the end of the chapter summarize the line types.

TRANSMISSION RATES

Perhaps the most important parameter in comparing communication lines is their transmission speed. The communication lines in use vary in transmission rate from 45 bits per second up to more than a million. Speeds higher than those of telephone lines, however, are not widely available on a dial-up basis, although switched high-bit-rate facilities now exist in some cities.

Communication lines are often categorized as three levels of transmission rate.

1. *Subvoice grade*—slower than telephone lines.
2. *Voice-grade* telephone lines.
3. *Wideband*—faster than telephone lines.

Subvoice-grade lines that transmit at rates from 45 to 180 bits per second are available in North America. In the rest of the world, 200 bits per second is an agreed upon international standard, and many countries have 200-bit-per-second lines.

Voice-grade lines used to be employed at speeds of 600 and 1200 bits per second, but more sophisticated modem designs have pushed the rate up to speeds as high as 10,500 bits per second. A penalty is paid for the very

high speeds in increased error rates and modem costs, plus the time it takes to reverse the direction of transmission. We will discuss these factors in the next chapter. Because of them, it is common to use transmission speeds over voice lines in the range 1200 to 4800 bits per second.

Wideband lines (also called *broadband*) are most commonly used at speeds of 19,200, 40,800 and 50,000 bits per second. Less commonly, speeds up to 500,000 bits per second are in use and higher bit rates are possible if required.

All these line types may be channeled over a variety of different physical facilities. This chapter, and indeed the tariffs themselves normally, say nothing about the medium used for transmission. It could equally well be wire, coaxial cable, microwave radio, or even satellite [1]. The transmission over different media is organized in such a way that the channels obtained have largely the same properties—same capacity, same noise level, and same error rate. The user generally cannot tell whether he is using a microwave link, coaxial cable, or pairs of open wires stretched between telephone poles. Only satellite transmission requires different data-handling equipment, and here only because a delay of one-third of a second or so is encountered in transmitting to the satellite and back.

SWITCHED VERSUS LEASED LINES

The next most important parameter about the lines is whether they are public switched lines or not. Voice lines and telegraph lines can be either switched through public exchanges (central offices), or permanently connected. Facilities for switching wideband channels are in operation in some countries, although most wideband channels today are permanent connections.

When you dial a friend and talk to him on the telephone, you speak over a line connected by means of the public exchanges. This line, referred to as a *public* or *switched* line, could be used for the transmission of data. Alternatively, a *private* or *leased* line could be connected permanently or semipermanently between the transmitting machines. The private line may be connected via the local switching office, but it would not be connected to the switchgear and signaling devices of that office. An interoffice private connection would use the same physical links as the switched circuits. It would not, however, have to carry the signaling that is needed on a switched line.

Just as you can either dial a telephone connection or have it permanently wired, so it is with other types of lines. Telegraph lines, for example, which have a much lower speed of transmission than is possible over voice lines, may be permanently connected, or they may be dialed like a telephone

line, using a switched public network. Telex is such a network; it exists throughout most of the world, permitting transmission at 50 bits per second. Telex users can set up international connections to other countries. Some countries have a switched public network, which operates at a somewhat higher speed than Telex but at less speed than telephone lines. In the United States the TWX network gives speeds up to 150 bits per second. TWX lines can be connected to Telex lines for overseas calls. Also, certain countries are building up a switched network for very high speed (wideband) connections. In the United States, Western Union has installed the first sections of a system in which a user can indicate in his dialing what capacity link he needs.

Many countries in the years ahead will be building data transmission networks of greater flexibility than existing facilities. They will have the ability to interconnect machines with links of widely varying speeds.

ADVANTAGES OF LEASED LINES

The great advantage of the public telephone network is its ubiquitousness. There are telephones everywhere, and wherever a telephone exists we can connect a data transmission machine to the line.

There are, however, some advantages in using leased lines that are permanently connected.

1. If it is to be used for more than a given number of hours per day, the leased line is less expensive than the switched line. If it is used for only half an hour per day, then it is more expensive. The breakeven point depends on the actual charges, which, in turn, depend on the mileage of the circuit, but it is likely to be of the order of an hour to several hours per day. This is clearly an important consideration in designing a data transmission network.
2. Private lines can be specially treated or "conditioned" to compensate for the distortion encountered on them. The common carriers charge extra for conditioning. In this way the number of data errors can be reduced or, alternatively, a higher transmission rate can be made possible. The switched connection cannot be conditioned beforehand in the same way, for it is not known what path the circuit will take. Dialing one time is likely to set up a quite different physical path from that obtained by dialing at another time, and there are a large number of possible paths. Modems now exist that condition dynamically and adjust to whatever connection they are used on. These devices enable higher speeds to be obtained over switched circuits but they are expensive.

3. Switched voice lines usually carry telephone company signaling within the bandwidth that would be used for data [2]. Data transmission machines must be designed so that the form in which the data are sent cannot interfere with the common carrier's signaling. With some machines, this also makes the capacity available for data transmission somewhat less than that over a private voice line. A common rate over a switched voice line in the 1960s was 1200 bits per second, whereas 2400 bits per second was common over a specially conditioned, leased voice line. Because of improved modem designs, it is probable that 3600 bits per second over switched voice lines and 9600 bits per second over conditioned, leased voice lines will become common in the 1970s. Already some modems transmit at higher speeds than 3600 over public voice lines.
4. The leased line may be less perturbed by noise and distortion than the switched line. The switching gear can cause impulse noise that results in data errors. This is a third factor that contributes to a lower error rate for a given transmission speed on private lines.

The cost advantage of switched lines will dominate if the terminal has only a low usage. Also, the ability to dial a distant machine gives great flexibility. Different machines, perhaps offering quite different facilities, can be dialed with the same terminal. A typewriter terminal used at one time by a secretary for computer-assisted text editing may at another time be connected to a scientific time-sharing system and at still another time may dial a computer-assisted teaching program. Machine availability is another consideration. If one system is overloaded or under repair, the terminal user might dial an alternative system. Often this dialing is done over the firm's own leased tie lines.

LINE CONDITIONING

As has been mentioned, private, leased voice lines can be *conditioned* so that they have better properties for data transmission [3]. Tariffs specify maximum levels for certain types of distortion. *An additional charge is made by most carriers for lines that are conditioned.*

The American Telephone and Telegraph Company, for example, has three types of conditioning in common use for voice lines carrying data, referred to as types C1, C2, and C4. A line ideal for data transmission would have an equal drop in signal voltage for all frequencies transmitted. Also, all frequencies would have the same propagation time. This is not so in practice. Different frequencies suffer different attenuation and different signal delay. Conditioning attempts to equalize the attenuation and delay at different frequencies. Standards are laid down in the tariffs for the measure of equalization that must be achieved. The signal attenuation and delay at different frequencies must lie within certain limits for the type C1, C2, and C4 conditioning [3]. The result of the conditioning is that a higher data speed can be obtained over that line, given suitable line-termination equipment (modems).

SIMPLEX, HALF-DUPLEX, AND FULL-DUPLEX LINES

In designing a data-processing system, it is necessary to decide whether that line must transmit in one direction only or in both directions. If the latter, will the machines transmit in both directions at the same time?

Transmission lines are classed as simplex, half duplex, and full duplex. In North America, these terms have the following meanings (Figure 4.1, page 39):

Simplex lines transmit in one direction only.

Half-duplex lines can transmit in either direction, but only in one direction at once.

Full-duplex lines transmit in both directions at the same time.

One full-duplex line is thus equivalent to two simplex or half-duplex lines used in opposite directions. The full-duplex line is sometimes referred to as, simply, *duplex*.

If the fireman in the last chapter was transmitting data over his hose in a *half-duplex* fashion, he would have to stop when he finished sending, reverse the direction of transmission and listen for the reply. This is true in data transmission also. There is a delay when the direction of transmission is reversed, called the *line turnaround time.*

For full-duplex transmission, the fireman would have used two hoses, one transmitting in each direction.

Simplex or half-duplex data transmission requires two wires to complete an electrical circuit. A full-duplex line can be thought of as being two simplex lines, one going in each direction. Usually a *four-wire circuit* is needed for full-duplex transmission. There is, however, an ingenious way to build what is in effect a four-circuit out of two wires. The bandwidth of the lines is split up into two separate frequency bands, one of which is used for transmission in one direction and the other of which transmits in the opposite direction. This is referred to as an *equivalent four-wire circuit* [4]. Although it is done with two wires, it works as though there were four wires of half the bandwidth, and permits full-duplex operation.

Data transmission machines often have specifications saying whether they require a two-wire or four-wire circuit.

The preceding meanings of the words "simplex" and "half-duplex" are in current usage throughout most of the world's computer industry. However, the International Telecommunications Union, an organization that has determined most of the world's standards in telecommunication, defines these terms differently.

Simplex (*circuit*). A circuit permitting the transmission of signals in either direction but not simultaneously.

Half-duplex (circuit). A circuit designed for duplex operation but which, because of the nature of the terminal equipment, can be operated alternately only.

Simplex and half-duplex are thus used differently by European telecommunications engineers and computer manufacturers (especially American ones) using such facilities.

Throughout this book the words will be used with the former meanings exclusively.

Public telephone lines are half duplex in operation. It is only with *leased* telephone lines that the user has a choice between half-duplex and full-duplex. In North America, full-duplex lines generally cost 10 percent more than half-duplex lines. In some countries they cost the same.

Simplex lines (American meaning) are not generally used in data transmission because even if the data are being sent in one direction, control signals are normally sent back to the transmitting machine to tell it that the receiving machine is ready or is receiving the data correctly. As a rule, error signals (positive or negative acknowledgement) are sent back so that there can be retransmission of messages damaged by communication line errors. Many data transmission links use half-duplex lines. This allows control signals to be sent and two-way "conversational" transmissions to occur. On some systems full-duplex lines can give more efficient use of the lines at little extra line cost. A full-duplex line often costs little more than a half-duplex line. Data transmission machines that can take full advantage of full-duplex lines are more expensive, however, than those that use half-duplex lines. Half-duplex transmission is, therefore, more common at present, although this situation might well change.

TARIFFS

The services offered by a telecommunication company to the public are described in tariffs. A tariff is a document that (in the United States) is required by the regulating bodies who control the carriers. It specifies details of the service and its cost. The United States Federal Communications Commission must eventually approve all interstate facilities, and similar state commissions control those within state boundaries. By law, all tariffs must be registered with these bodies. In most other countries, the telecommunication facilities are set up by government bodies and thus are directly under their control.

We will now discuss the tariffs that are of use for data transmission. (These are summarized in Table 4.2.)

TELEX

Telex is a worldwide switched, public teleprinter system. It operates at 66 words per minute (50 bits per second) and uses a 5-bit per character code described in Chapter 6. It is operated in the United States by Western Union. Any teleprinter on the system can dial any other teleprinter in that country, and Telex machines can be connected internationally without speed or code conversion. The United States can dial Canada and Mexico directly, but operator intervention is needed to other countries. Some countries permit the Telex facilities to be used for other forms of dial-up data transmission. Each Telex call is billed on a time-and-distance basis.

Each subscriber has an individual line and his own number, as with the conventional telephone service. His teleprinter is fitted with a dial, like a telephone, with which he can dial other subscribers. The teleprinter used may or may not have paper-tape equipment also. The teleprinter can be unattended. When a message is sent to an unattended teleprinter, it will switch itself on, print the message, and then switch itself off.

TELETYPEWRITER EXCHANGE SERVICE

The North American common carriers offer a service that is competitive with Telex. Again each subscriber has a dial-up teletypewriter with his own number listed in a nationwide directory. This service is called the Teletypewriter Exchange Service (TWX).

Other manufacturers' data transmission equipment can be connected to TWX lines and can transmit at speeds up to 150 bits per second, half or full duplex. This process requires a special terminal arrangement at additional cost. There are three types of access lines to thc TWX network, as follows:

1. TTY-TWX.
   This is an access line with a teletypewriter provided by the common carrier. The speeds of transmission are either six characters per second in Baudot code or ten characters per second in Data Interchange Code (DIC).
2. CPT-TWX.
   CPT stands for "customer-provided terminal"; to this access line the customer can attach any device operating with one of the above two speeds and codes and adhering to normal TWX line control. The device could be a computer with an appropriate adapter on its input-output channel.
3. CE-TWX (formerly called "TWX Prime").
   CE stands for "customer equipment." This can now be any device and is not restricted to a specific code or character speed. Two TWX subsystems are accessible, one that operates at speeds up to 45 bits per second, and the other up to 150. A CE-TWX terminal can communicate only with another CE-TWX terminal.

TWX directories are published listing TTY-TWX and CPT-TWX subscribers, but not CE-TWX subscribers as they cannot necessarily be interconnected.

WIDE AREA TELEPHONE SERVICE

A further tariff in the United States provides dial-up facilities for a fixed monthly fee and is called WATS, the Wide Area Telephone Service. When a subscriber is given an (outward) WATS access line, he is permitted to make as many calls as desired to a specified zone. Subscribers in that zone are similarly permitted to call an (inward) WATS line. These would be two separate lines. On either there are two alternative methods of charging, one using a flat monthly charge and the other using a fixed hourly rate with a minimum of 10 to 15 hours.

Suppose that a subscriber in New York wishes to have a WATS line to the West Coast of North America. He may pay a certain fixed charge per month for this line and is then free to make whatever calls he wants to the West Coast or nearer WATS zones. The whole of the United States is divided up into six WATS zones based largely on state boundaries. Figure 4.2 shows these zones as viewed from New York. The charge for zone 2 is greater than that for zone 1, zone 3 greater than 2, and so on. If the subscriber pays for access to zone 4, he can call zones 1, 2, and 3 for that same charge, but not zones 5 and 6.

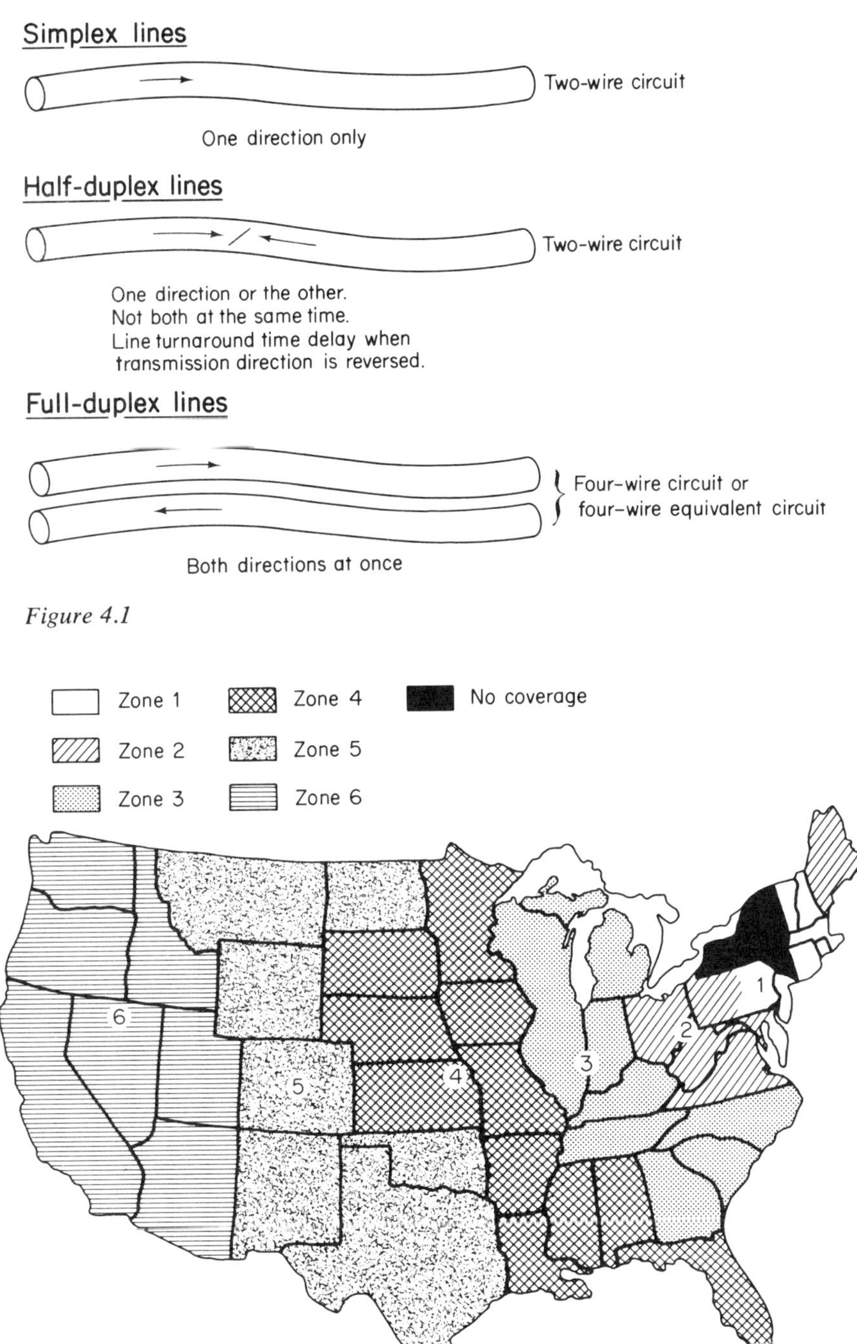

Figure 4.1

Figure 4.2. WATS zone configuration for southeast New York subscriber.

The zone numbering of Figure 4.2 is applicable only to New York. Another state would have a map with different numbering and different charges. Figure 4.3 shows the WATS zone numbering as applicable to Missouri.

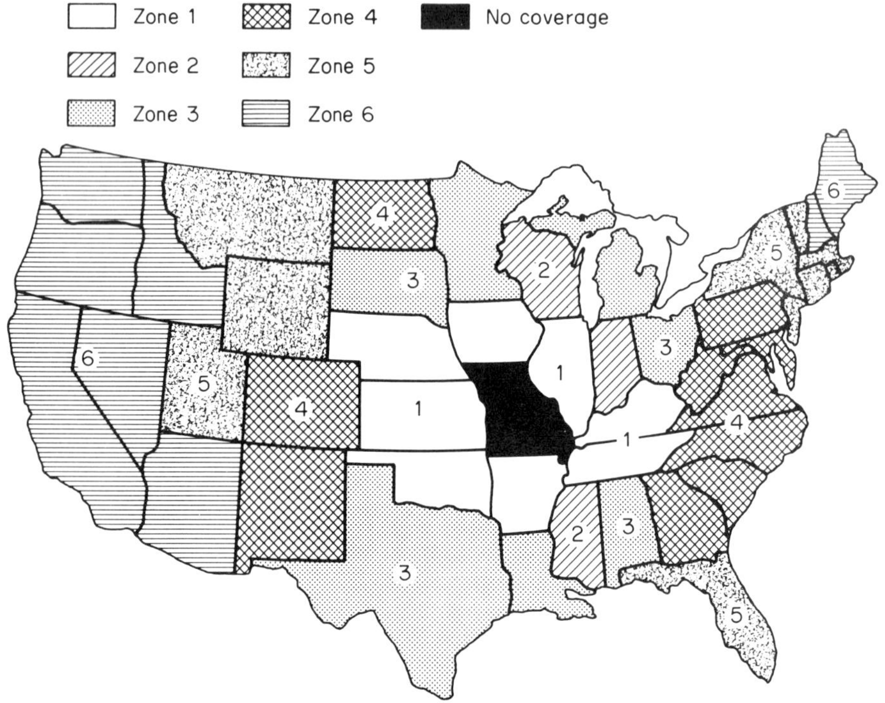

Figure 4.3. WATS zone configuration for a Missouri subscriber.

The two types of WATS service are called OUTWATS, in which the subscriber with the WATS line may originate as many calls as he wishes to the zone in question, and INWATS, in which subscribers in the zone may call the WATS line. INWATS and OUTWATS require separate lines. Both are used to help minimize data transmission costs.

WATS permits transmission using the same data sets as normal telephone lines. It is not possible to make person-to-person, conference, third party, credit, or collect (reverse charge) calls under the WATS tariff.

TARIFFS FOR WIDEBAND LINES AND BUNDLES

The North American common carriers offer several tariffs for leased wideband lines. Some of these can be subdivided by the carrier into "bundles" of lower bandwidth. Some can be subdivided into channels for voice transmission, telephotograph, teletypewriter, control, signaling, facsimile, or data. With some tariffs the user pays a lower price for the bundles than for the individual channels.

The word "TELPAK" was formally used for the "bulk" communication services offered by the telephone companies and Western Union. The word has now been eliminated from the tariffs but is still found in much literature. What used to be TELPAK C is now called a Type 5700 line, and what used to be TELPAK D is now Type 5800. Both can provide a wideband channel or a bundle of lesser channels.

Originally there were four sizes of TELPAK channels: TELPAK A, B, C, and D. However, in 1964 the Federal Communications Commission ruled that rates for TELPAK A (12 voice circuits) and TELPAK B (24 voice circuits) were discriminatory in that a large user could obtain a group of channels at lower cost per channel than a small user, who could not take advantage of the bulk rates. In 1967 the TELPAK A and B offerings were eliminated.

A Type 5700 line has a base capacity of 60 voice channels (full duplex).

A Type 5800 line has a base capacity of 240 voice channels (full duplex).

Each voice channel in these lines can itself be subdivided into one of the following:

1. Twelve teletype channels, half or full duplex (75 bits per second).
2. Six Class D channels, half or full duplex (180 bits per second).
3. Four AT & T Type 1006 channels, half or full duplex (150 bits per second).

There cannot be mixtures of these channel types in a voice channel. The Type 5700 line can transmit data at speeds up to 230,400 bits per second; the Type 5800 line has a potential transmission rate much higher. Line-termination equipment is provided with these links, and each link has a separate voice channel for coordination purposes.

The TELPAK channels thus serve two purposes. First, they provide a wideband channel over which data can be sent at a much higher rate than over a voice channel. Second, they provide a means of offering groups of voice or subvoice lines at reduced rates—a kind of discount for bulk buying.

Suppose that a company requires a 50,000-bit-per-second link between two cities, together with 23 voice channels and 14 teletypewriter channels; or perhaps 30 voice channels and no teletypewriter link. Then it would be likely to use the Type 5700 tariff. In leasing these facilities, it would have some unused capacity. If it wishes, it can make use of this at no extra charge for mileage, although there would be a terminal charge.

Government agencies and certain firms in the same business whose rates and charges are regulated by the government (e.g., airlines and railroads) may share bundled services. Airlines, for example, pool their needs for voice and teletypewriter channels. An intercompany organization purchases the bundled services and then apportions the channels to individual airlines.

Series 8000 is another "bulk" communications service in the United States that offers wideband transmission of high-speed data, or facsimile, at rates up to 50,000 bits a second; the customer has the alternative of using the channel for voice communication up to a maximum of 12 circuits. A Type 8801 link, part of this series, provides a data link at speeds up to 50,000 bits per second with appropriate terminating data sets and a voice channel for coordination. A Type 8803 link provides a data link with a fixed speed of 19,200 bits per second and leaves a remaining capacity that can be used either for a second simultaneous 19,200-bits-per-second channel or for up to five voice channels. These links must connect only two cities. The separate channels cannot terminate at intermediate locations.

Most countries outside North America also offer tariffs similar to the Series 8000, and in most locations quotations for higher speeds can be obtained on request. Obtaining a wideband link in many such countries can be a slow process. This is particularly so if the termination is required in a small town or rural area rather than in a city to which such links already exist. No doubt, as the demand for such facilities increases, so the service of the common carriers in providing them will improve.

SWITCHED WIDEBAND SERVICES

TELPAK and the Series 8000 lines are not switched. They can, however, be linked to private switching exchanges within a corporation, to provide a corporate switched wideband network.

Two public wideband networks are in operation in North America:

1. *Dataphone 50*.
   An AT & T service offering switched wideband lines that transmit data at 50,000 bits per second. So far it is available in only a few cities.
2. *BEX (Broadband Exchange System)*.
   Western Union network in which the user includes a digit in his dialing to indicate what bandwidth he requires. He will eventually be able to transmit at 600, 1200, 2400, 4800, 9600, and 38,400 bits

per second. The higher speeds are not yet available and the network serves only certain cities.

SWITCHED PRIVATE SYSTEMS

Many firms have private-leased-line systems which are switched with private exchanges. It is possible to engineer these systems to the same quality as private lines and hence provide a switching system of better quality than the public network.

Some private lines are wholly owned by their users rather than leased. Users are generally prohibited from installing their own lines across public highways, and most privately installed communication links are wholly within a user's premises—for example, within a factory, office building, or laboratory. Railroads have their own communication links along their tracks. Some companies have private point-to-point microwave transmission links or other radio links.

Optical and infrared links have been used for short-distance high-data-rate links between buildings. The main drawback of such links is that they are adversely affected by immense storms.

Millimeterwave radio—a little higher in the radio spectrum than microwave—is now used effectively to give short-distance links carrying more than a million bits per second. The millimeterwave antennas are small, easy to install, fairly inexpensive, and their user does not require a license.

SUMMARY

A summary of the main categories of communication links is given in Tables 4.1 and 4.2.

Table 4.1. CATEGORIES OF COMMUNICATION LINE AND TARIFF

Types of Link	Comments
Digital link	Designed for digital transmission. No modem required. Are code-sensitive in some cases.
Analog link	Transmits a continuous range of frequencies like a voice line. Modem required.
Switched public	Cheaper if usage is low. Switched telephone lines are universally available.
Leased (sometimes called "private")	Cheaper than public lines if usage is high. May have lower error rate. Higher speeds possible on leased telephone lines than switched ones.
Leased with private switching	May give the lowest cost. Combines the advantages of leased lines with the flexibility of switching. Public switched wideband lines may not be available.
Private (noncommon-carrier)	Usually only permitted within a subscriber's premises. See next item.
Private (noncommon-carrier) links:	
In-plant	Very high bit-rates achievable.
Microwave radio	Permissible in special cases for point-to-point links.
Shortwave or VHF radio	Used for transmission to and from moving vehicles or people.
Optical or infrared	Used for short links—e.g., intercity—at high bit rates (250,000 bps, typical). No license required. Put out of action by fog or *very* intense rain.
Speeds	
Subvoice grade	Usually refers to speeds below 600 bits per second.
Voice grade	Usually refers to analog voice lines using modems of speeds from 600 to 10,500 bits per second.
Wideband	Speeds above those of voice lines, most commonly 19,200; 40,800; 50,000, and 240,000.
For a detailed list, see the following table.	
Mode of operation	
Simplex	Not normally used in data transmission.
Half duplex	The most common in the United States.
Full duplex	On the Bell System costs 10% more than half duplex and can give disproportionately higher throughput if the terminal is designed to take advantage of it.

NOTE: These terms sometimes describe the limitation of a machine rather than the limitation of the line it is attached to.

Table 4.1 Continued

Types of Link	Comments
Conditioning Types: C1, C2, C4	Applied to leased analog voice lines to achieve more uniform attenuation and delay so that higher-speed modems can be employed.
Tariffs Tariff includes data set	Common carrier provides the data set (modem) and hence the transmission rate is fixed. (Explained in following chapter).
Tariff simply for bandwidth	User chooses the modem. Transmission rate depends on modem design.
DATA-PHONE service	An AT & T service that provides a data set with the line.
Public telegraph networks: Telex	Worldwide telegraph netword, Baudot code, 6.7 characters per second.
TWX	North American telegraph network. Baudot code at 6 characters per second or DIC code at 10 characters per second.
TWX: TTY-TWX	With common carrier teletype machines.
CPT-TWX	With customer-provided terminals adhering to TTY-TWX codes and line control.
CE-TWX	With customer equipment. Need not adhere to TTY-TWX codes. Speeds up to 150 bits per second. Not listed in TWX directories.
WATS (Wide Area Telephone Service) WATS: INWATS	Any subscriber in the specified zone may call *to* the WATS line.
OUTWATS	The call originator is connected to the WATS line and may call any subscriber in the specified zone.
	INWATS or OUTWATS must be on separate lines.
TELPAK	A form of quantity discount in which a bundle of lines can be purchased.
Bundling: TELPAK C, Type 5700	Equivalent to 60 voice channels.
TELPAK D, Type 5800	Equivalent to 240 voice channels. The TELPAK bandwidth can be subdivided in many different ways.

Table 4.2. TYPES OF COMMUNICATION LINES AVAILABLE

1. *Public* (*Dial-up*) *Lines*	Bit Rate (bits per second)		Bandwidth (kHz)	Type of Line			Half Duplex or Full Duplex
	Fixed	Dependent on Modem		United States: AT & T	United States: Western Union	United Kingdom	
Subvoice grade	45				TTY-TWX and CPT-TWX		HDX
		Up to 45			CE-TWX		HDX or FDX
	50				Telex	Telex	HDX
	110				TTY-TWX and CPT-TWX		HDX
		Up to 150			CE-TWX		HDX or FDX
		Up to 200				Datel 200	FDX
The public telephone network		Up to 600	3	Public network		Datel 600	FDX
		Up to 1200	(Not all freely usable because of network signaling.)			Datel 600	FDX
		600 to 9600					HDX or FDX
Switched wideband networks	600 1200 & 2400 4800* 9600* 38,400*	(Other speeds will be achievable with other modems.)	2 4 8* 16* 48*		BEX (certain cities only)		FDX
	50,000	Up to 50,000		Dataphone 50 (few cities only)			FDX

*Planned but not yet available.

Table 4.2 Continued

2. Leased Lines

Subvoice grade		Up to 45		1004			HDX/FDX
	50					Tariff H	FDX
		Up to 55		1002			HDX/FDX
		Up to 75		1005			HDX/FDX
	100					Datel 100	FDX
		Up to 150		1006			HDX/FDX
		Up to 180			1006		HDX/FDX
	200					Datel 200	FDX
Voice-grade lines (For the higher rates, conditioning is needed.)		Up to 600	3			Datel 600	FDX
		Up to 1200	3			Datel 600	FDX
		600 to 10,500	3	3002 (C1, C2, and C4 conditioning)	3002	(Datel 2000 refers to conditioned or high-quality voice lines)	HDX/FDX
Wideband	19,200		24	8803	8803	Special quotation	FDX
	40,800		48	8801	8801		FDX
	50,000		48	8801	8801		
	230,400		240	5700 or [Originally called TELPAK C and D (see text), can be used as "bundles" of small bandwidth lines.]	5800		FDX

5 MODEMS AND DATA SETS

Telephone lines were designed for transmitting the sound of the human voice—not all of it, as we mentioned two chapters ago; enough of it to make it intelligible and recognizable. Nobody sings opera over the telephone.

Data leaving a terminal in no way resemble the human voice. In order to transmit such information, we manipulate it electronically until it *does* occupy the same range of frequencies as the telephone voice. This is the function of a *modem*. The process is called *modulation*. At the other end of the line, a similar device must carry out the converse process of converting the transmitted signal back into pulses that mean something to the receiving machine. This is called *demodulation*. The same little box of electronics does both the *modulation* and *demodulation*. The word modem is a contraction of these words. A modem is called a "data set" by AT & T, and may include the telephone dial and hand set. It is sometimes called a "line adaptor" by IBM.

Figure 5.2 shows a typical modem and Figure 5.1 illustrates its use.

In addition to converting the signal into a form in which a high bit rate can be sent, the modem serves to protect the lines from undesirable signals that might cause interference with other users or with the network's signaling system.

A variety of different modulation processes are possible. There is much scope for ingenuity in the design of modems, and the increasing speed at which data can be sent over the telephone lines is largely due to improving modem design. Often the data-processing system designer is faced with a choice of modems, each having different characteristics. Some will be inexpensive but slow. Some will be fast but error prone. Some will be accurate,

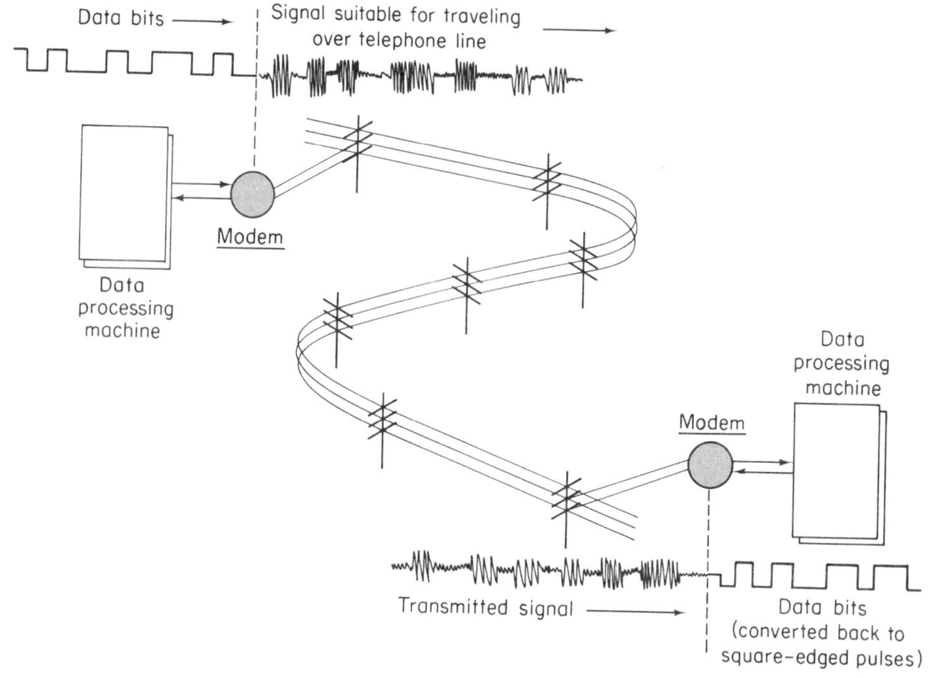

Figure 5.1. Use of modems.

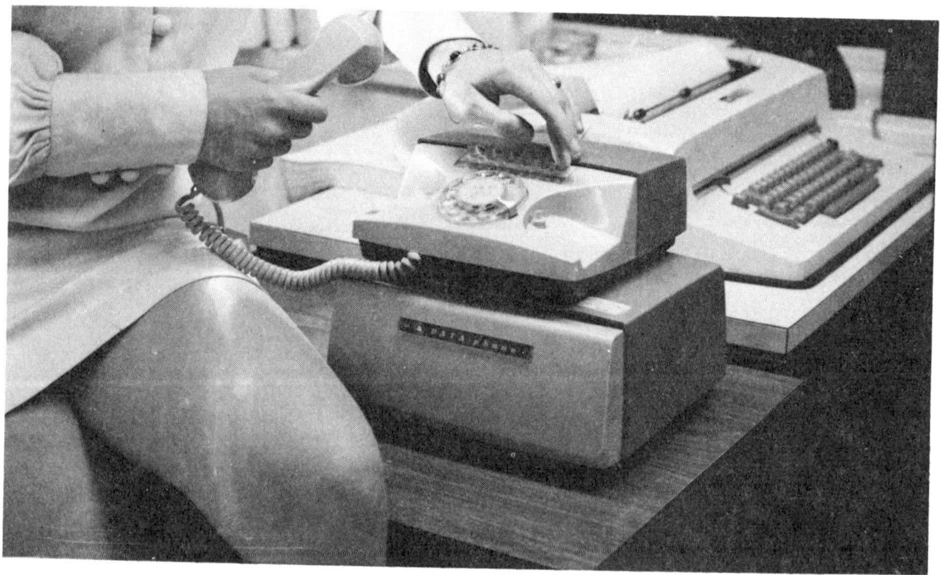

Figure 5.2. An operator dialing a connection with a typical Bell System modem (data set).

expensive and moderately fast. Some will have a long turnaround time. Different systems differ in their requirements, and the designer can choose his modems accordingly.

NO MODEM

The simplest (and cheapest) arrangement of all is not to have a modem. The square-edged pulses that emerge from the terminal are fed directly into the pair of wires that are used for transmission. This is referred to as *baseband* signaling. The pulses quickly become distorted, as did the pulses transmitted down a firehose and shown in Figures 3.2 and 3.3. If they are transmitted too far or too fast, the distortion

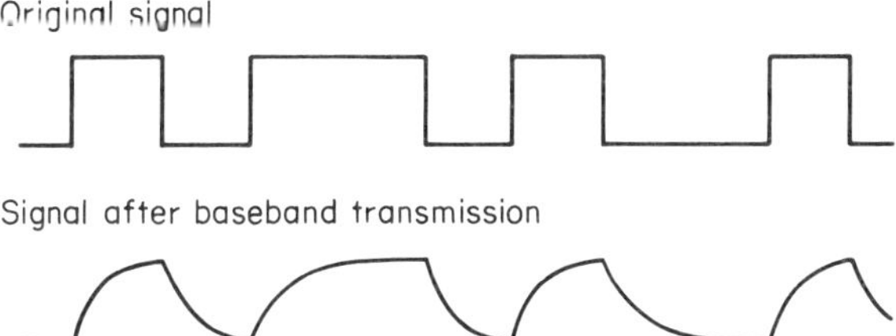

Figure 5.3. When the bit pattern is transmitted with modems, the pulses become distorted as shown. If they are transmitted too far or at too high a bit rate, the distortion will be so great that the original bit pattern cannot be recovered.

will become so great that the bit pattern cannot be recovered. There is one other problem: they cannot be transmitted over a line with amplifiers because these devices will not amplify the dc component of baseband signals.

However, this form of transmission is perfectly satisfactory for local operation. Terminals operating at up to 300 bits per second can operate successfully over lines without amplifiers up to 3 miles. Much higher speeds are possible over shorter distances. The "local loop" from the subscriber to the telephone central office does not have amplifiers if it is only a few miles in length. Baseband signaling cannot be used, in general, over the public network because of the longer distances and amplifiers.

ACOUSTICAL COUPLERS

When bits are sent too far or too fast for baseband signaling, they must be converted to appropriate frequencies for the transmission line in question.

One inexpensive method is to use acoustical couplers. These devices convert the characters to be transmitted into audible tones, such as those generated by a Touchtone telephone. A machine using audible tones need not necessarily be physically wired to the line. It can make sounds that are picked up directly by the telephone handset. This process is referred to as acoustical coupling. The sounds used must be at those frequencies well within the telephone bandwidth. The sounds are reproduced at the far end of the connection by a telephone earpiece. They are then converted back into data signals. Acoustical coupling is somewhat less efficient than direct coupling. At the time of writing it is used for transmitting between relatively slow machines, such as typewriterlike terminals.

In an acoustical-coupling device, the telephone handpiece may fit into a special cradle, as shown in Figure 5.4. Acoustical couplers are generally less expensive than modems. Another advantage of acoustical coupling is that the terminal can easily be made portable. A small terminal could be made to transmit to a computer from a public call box. Nondigital machines also use acoustical coupling. Documents can be copied at a distance with a Xerox machine in this way.

Although there is no electrical connection to the telephone lines, it is still possible for acoustically coupled machines to interfere with the public network's signaling, as with a directly coupled device. Also, severe cross talk can be caused on a telephone link that carries many voice channels, by the transmission of a continuous frequency, as with a repetitive data pattern. The coupling device must be designed to avoid these difficulties.

By 1970 acoustical couplers were extensively used in the United States. In many other countries, however, the telecommunication authorities would not permit their use. The result was a prohibition of the various applications of small portable terminals, such as from public coin-operated telephones. A

Figure 5.4. Acoustical coupler.

major advantage of acoustical couplers is being able to use them anywhere there is a telephone, without a wired connection to the telephone line. For this reason, acoustical couplers have been built into the circuitry of some portable devices. There are many possible ways of coding the characters into sets of audible frequencies that can be used on the telephone line.

DIRECT TONE TRANSMISSION

The same coding and frequencies used with an acoustical coupler could be used with the frequency-generating circuits wired directly to the telephone line. This connection would be cheaper to manufacture than an acoustical coupler, for the telephone handset cradle with its microphone, speaker, and sound insulation would not be required.

Some inexpensive modems operate in this way, with data being sent character by character, and the bits in each character being sent in parallel with appropriate coding into frequencies. Such a technique, while inexpensive, cannot achieve a high character rate. It is satisfactory for single typewriter-speed devices on a voice line. Error-detection techniques can be used to guard against errors, but a higher error rate tends to be encountered than with other low-speed forms of modulation.

MODULATION OF A SINE WAVE CARRIER

There are two main purposes in using modems. First, they increase the possible speed over a given circuit, such as a voice line. Second, they reduce the effects of noise and distortion. Many communication links would be unusable at reasonable speeds without modulation. A variety of modems has been developed in recent years for use with data-processing machines. It is an area that is still developing fast.

The majority of modems in operation transmit a continuous sine wave (Figure 5.5) and then modify it in accordance with the data that are to be sent. The sine wave carrier may be represented as a function of time, t, by the equation

$$a = A \sin (2\pi ft + \theta)$$

where a = the instantaneous amplitude of carrier voltage at time t
A = the maximum amplitude of carrier voltage
f = the carrier frequency
θ = the phase

The values of A, f, or θ may be varied to make the wave carry information. There are therefore three basic types of modulation. They are called amplitude modulation, frequency modulation, and phase modulation.

Figure 5.6 illustrates the three types. A sinusoidal carrier wave of, say,

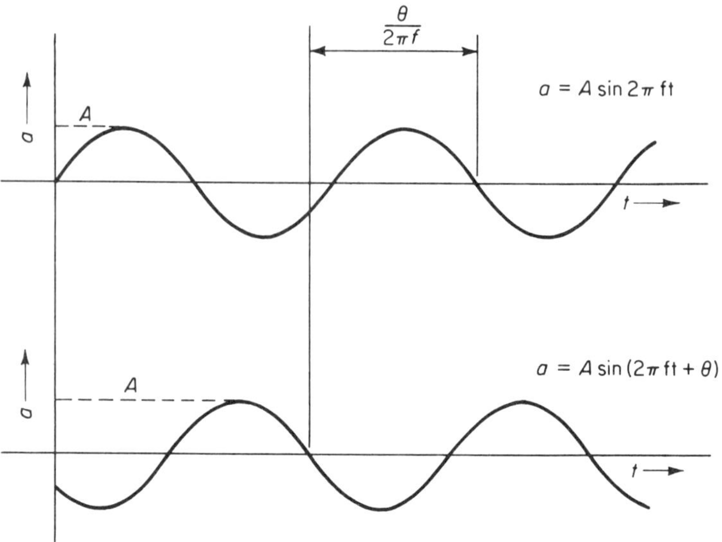

Figure 5.5. Two sine waves of frequency f with phase difference θ.

1500 hertz—in the center of the telephone voice band—is modulated to carry the information bits 01000101100.

In the top diagram of Figure 5.6, the *amplitude* is varied in accordance with the bit pattern. In the middle diagram, the *frequency* is varied; and in the bottom one, the *phase* is varied. In these simplified diagrams, the channel is being operated inefficiently because far more bits could be packed into the carrier oscillations shown. The tightness of this packing determines the speed of operation.

Many variations are possible within these three main types of modulation [1]. With frequency modulation, for example, the maximum frequency deviation that can occur can be varied over a wide range. The larger the deviation, the larger the bandwidth required but the greater the resiliance to the effects of noise. Different methods are possible for detecting the signal and recovering the bits. With phase modulation, for example, a given phase condition may represent either a given bit condition or a *change* from a 1 to 0 or 0 to 1 bit. The former is referred to as *fixed-reference detection* and the latter as *differential detection.*

MULTILEVEL TRANSMISSION

The previous figures all relate to binary transmission. It is possible, however, to use a waveform with *more than two states* as the modulating waveform. This is shown for amplitude modulation in Figure 5.7. *Di-bits* (i.e., pairs of bits) are transmitted rather than bits and need four possible states in the modu-

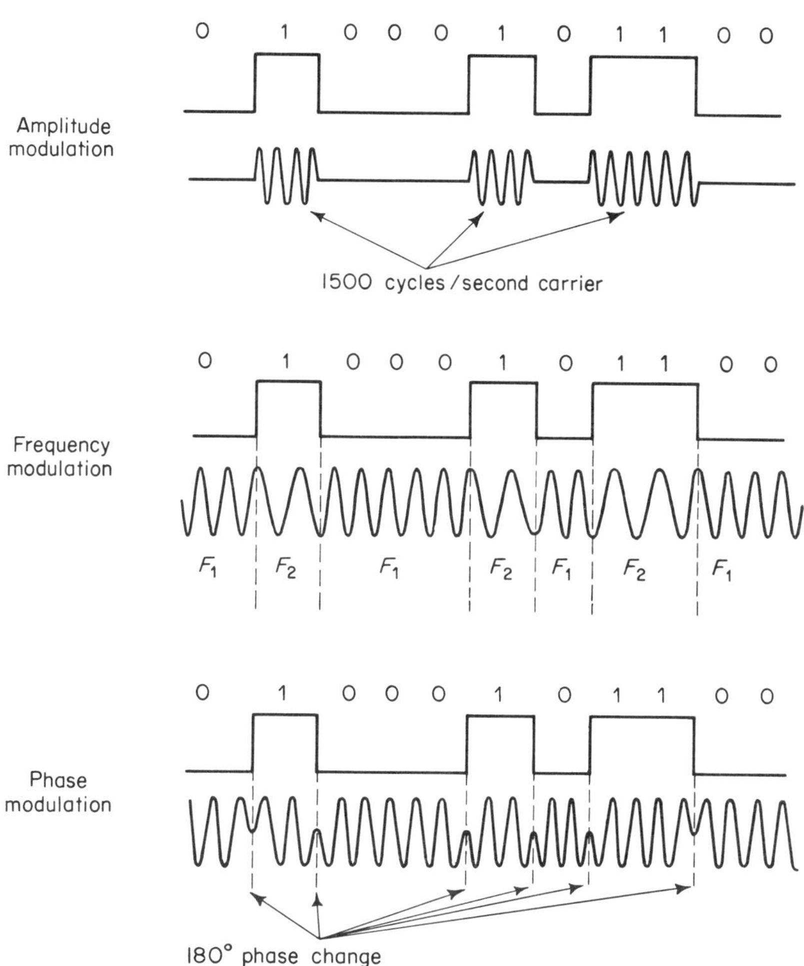

Figure 5.6. The three basic methods of modulating a sine wave carrier (a simplified diagram showing only binary signals).

lated waveform. This approximately doubles the transmission rate that can be achieved with the modem, but also considerably increases its susceptibility to noise. Because amplitude modulation is already susceptible to noise, di-bits are not normally used with this form of modulation; however, they are frequently used with phase modulation.

In some modems, *tri-bits* are used. To transmit the grouping of three bits, $2^3 = 8$ possible states are needed. Again, the susceptibility to noise and distortion is much greater. Phase modulation using di-bits is referred to as *four-phase modulation* and that using tri-bits is called *eight-phase modulation.*

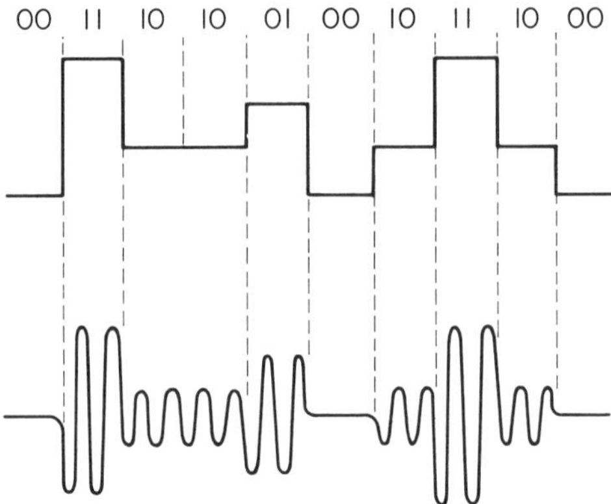

Figure 5.7. Di-bits are encoded here, rather than single bits, and hence there must be four detectable signal states instead of two. Similarly, tri-bits can be encoded with eight signal states.

BAUDS

The speed of a transmission line is often quoted as being a certain number of "bauds," which means *the number of times the line condition changes per second.* If the line condition represents the presence or absence of one bit, then the signaling speed in bauds is the same as bits per second. If, however, the line can be in one of four possible states at any instant as in Figure 5.7, then one line condition represents a "di-bit." x bauds will then be the same as $2x$ bits per second. When the signals are coded into eight possible states, one line condition represents three bits. One baud then equals three bits per second, and so on.

The reader should note that the term "bauds" is sometimes taken to mean "bits per second." Although this meaning is true with lines that use two-state signaling, it is not true in general. For any line not using two-state signaling, it is wrong. *The term* "bauds" *has therefore sometimes proven confusing and will be generally avoided throughout this book.*

LINE CONDITIONING

When four-state signaling is used, as in Figure 5.7, the demodulation device must detect smaller signal differences than when two-state signaling is used, as in Figure 5.6. When eight-state signaling is used, still smaller differences must be distinguished. This means that eight-state signaling is more likely to make errors in the presence of noise and distortion on the line. Other techniques for increasing the transmission speed also tend to increase the susceptibility to error.

For the highest-speed modems, it is necessary to control the distortion on the line as far as possible. Two types of distortion can be controlled. The first is called *attenuation distortion.* The signal is attenuated differently at different frequencies. The second is called *delay distortion.* The signal at different frequencies arrives at slightly different times—that is, there is a different delay at different frequencies.

Both attentuation distortion and delay distortion can be compensated for by carefully adjusting the circuit. Compensating adjustments can, in fact, be made at the ends of the circuit where the modems are connected.

When a public circuit is dialed between two distant locations, different physical paths will be switched into the links on different occasions. A call dialed from New York to Miami might go via Atlanta at one time and via Houston at another. It might even go via Los Angeles if the circuits in the Eastern United States are congested. You never can tell. With a private line, however, the path is fixed and adjustments can be made to its characteristics. In particular, it can be adjusted to have the same attenuation and the same delay, within certain limits, over the range of frequencies of the telephone voice. The attenuation and the delay are said to be "equalized" over the frequency range in question. *Equalizer* circuits achieve this by adding an artificial delay at different frequencies. The entire process can be performed by the telephone company, and this is what is referred to as "conditioning."

Many carriers either have separate tariffs for conditioned telephone lines and unconditioned lines or charge extra for conditioning, as was discussed in the previous chapter. Several grades of conditioning are available. A modem designed for a conditioned line will give a higher bit rate than that for an unconditioned line, or else will give a lower error rate.

MODEMS FOR PUBLIC LINES

On a private leased telephone line, the machine transmitting data has the whole telephone bandwidth to itself. On the public network, however, certain frequencies are used by the network itself for signaling purposes. The modem must be designed so that it cannot interfere with the signaling [2].

Many data-processing machines are designed to avoid the signaling frequencies completely. This limits the bandwidth available for transmission and hence the speed at which data can be sent. It is possible to design a modem that randomizes the signal, smearing it across the available bandwidth like the human voice so that it can never be mistaken for line signaling. However, doing so increases the modem cost.

Because of the signaling frequencies, *modems designed for private voice lines cannot operate on public lines.* The maximum speed attainable on public lines is lower. Furthermore, modems designed for the public network in North America cannot be used in other countries, and vice versa.

Furthermore, public lines cannot be *conditioned* as can private lines, because the path the connection will be made over differs. One time you dial you will obtain one path and another time a different path with different attenuation distortion and delay distortion.

Private lines thus not only have the advantage of being free from net-

work signaling but also are capable of being conditioned. In addition, they are generally less perturbed by noise and distortion than the switched line. The switching gear can cause impulse noise that causes errors in data. Because of the preceding factors, the data transmission speeds available on private voice lines have generally been twice those on public lines.

The designers of modems for public lines cannot benefit from conditioning. They can, however, fight back with what is referred to as "dynamic equalization." This process refers to a circuit that measures the characteristics of the line in use and automatically adjusts the attenuation and delay so that these factors are equal, within limits, over a given range of frequencies. In this way higher speeds can be obtained over public lines, but the cost of the modems is substantially increased.

THE DESIGNER'S CHOICE

These and other factors in the modem design give the system planner a range of possibilities in modem selection. He needs to evaluate the properties of the different modems in terms of what he intends to accomplish with the communication line network [3].

INTERFACE BETWEEN MODEMS AND MACHINES

It is very important that the interface between the modem and the various data transmission machines be standardized, so that one modem can be replaced by another, and so terminals and transmission control units can be interchanged. A standard interface between modems and business machines was specified by the Electronics Industries Association, and in practice is closely adhered to. It is the EIA RS-232-C Standard, which is illustrated in Figure 5.8. The international CCITT V. 24 interface is functionally similar, though differs in some details.

The standard interconnections between the modem and terminal (or any other business machine) are shown in Fig. 5.8. There are four types.

1. ***Data Signals.***
   (a) TRANSMITTED DATA (*to the modem*). Data generated by the terminal for transmission.
   (b) RECEIVED DATA (*to the terminal*). Data received by the modem for the terminal.
2. ***Timing Signals.***
   (a) TRANSMITTER SIGNAL ELEMENT TIMING. Two connections are defined. One sends signal element timing information from the transmitting terminal to its modem. The other sends timing information from the transmitting modem to its terminal.

a

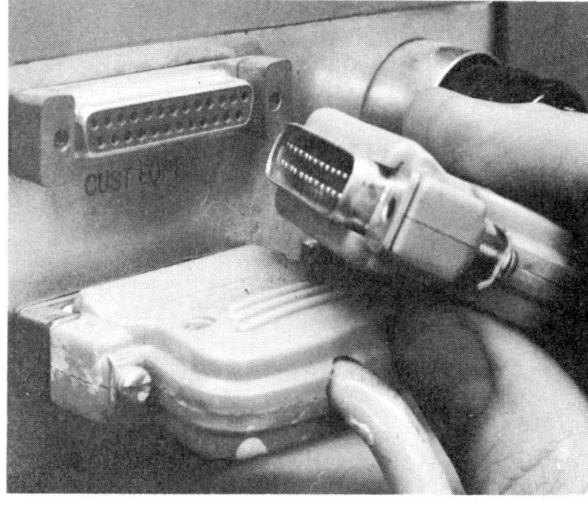

Fig. 5.8. This is the 25-pin plug between the modern and the data processing equipment. The interface between the modem and the data processing equipment is a standard that is well adhered to.

b

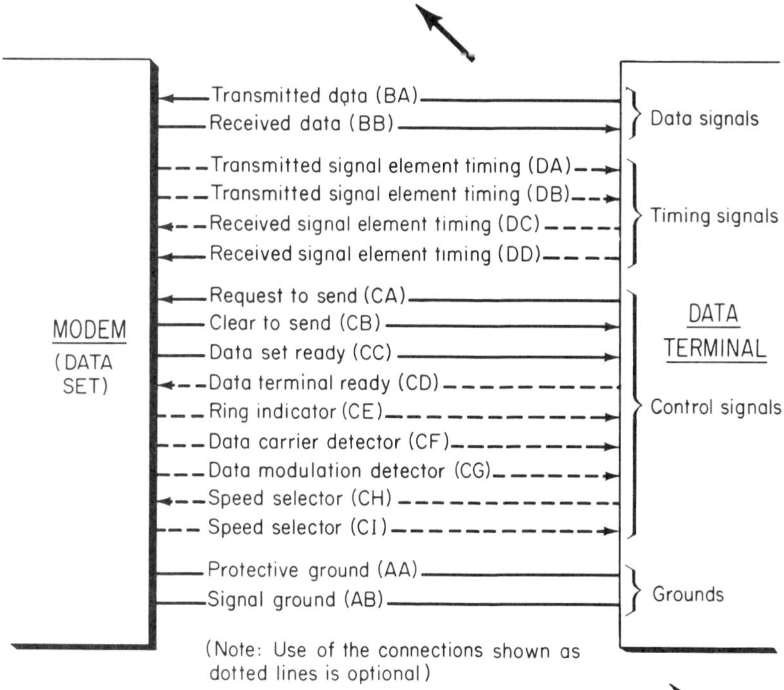

MODEM
(DATA SET)

Transmitted data (BA)
Received data (BB)

Transmitted signal element timing (DA)
Transmitted signal element timing (DB)
Received signal element timing (DC)
Received signal element timing (DD)

Request to send (CA)
Clear to send (CB)
Data set ready (CC)
Data terminal ready (CD)
Ring indicator (CE)
Data carrier detector (CF)
Data modulation detector (CG)
Speed selector (CH)
Speed selector (CI)

Protective ground (AA)
Signal ground (AB)

Data signals
Timing signals
DATA TERMINAL
Control signals
Grounds

(Note: Use of the connections shown as dotted lines is optional)

c

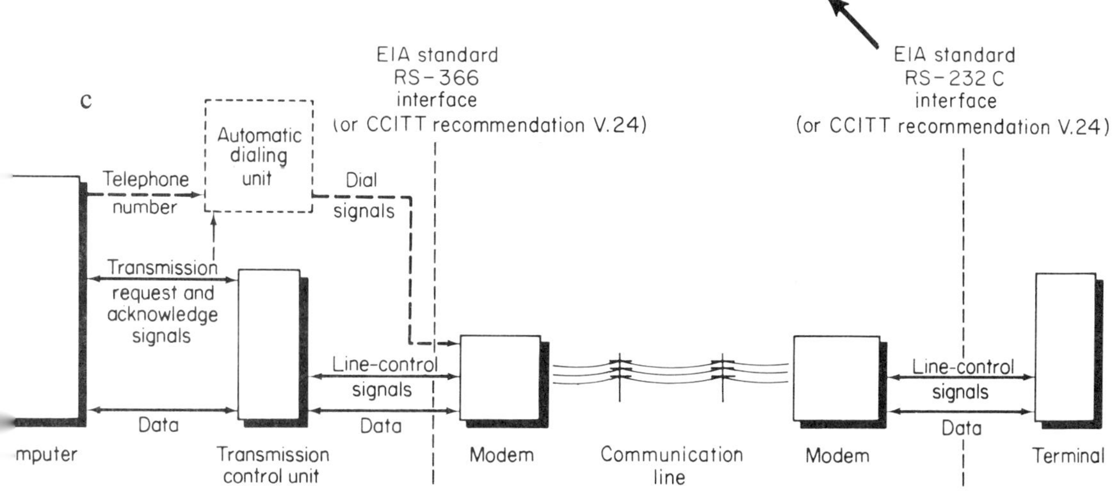

EIA standard RS-366 interface (or CCITT recommendation V.24)

EIA standard RS-232 C interface (or CCITT recommendation V.24)

Telephone number
Automatic dialing unit
Dial signals
Transmission request and acknowledge signals
Line-control signals
Data
Data
Line-control signals
Data

mputer
Transmission control unit
Modem
Communication line
Modem
Terminal

(b) RECEIVER SIGNAL ELEMENT TIMING. Two connections are defined. One sends signal element timing information from the *receiving* terminal to its modem. The other sends timing information from the receiving modem to its terminal.

The timing signal connections are optional. A modem for START–STOP transmission does not use them.

3. ***Control Signals.***

(a) REQUEST TO SEND (*to the modem*). Signals on this connection are generated by the transmitting terminal when it wishes to transmit. The modem's carrier signal is transmitted during the ON condition of this connection. (With half-duplex operation, the OFF condition of this connection holds the modem in the receive-data state.)

(b) CLEAR TO SEND (*to the terminal*). Signals on this connection are generated by the transmitting modem to indicate that it is prepared to transmit data. They are a response to the REQUEST TO SEND signal from the transmitting device. (With full-duplex operation the modem is in the transmit state at all times.)

(c) DATA SET READY (*to the terminal*). Signals on this connection are generated by the local modem to indicate to the transmitting machine that it is ready to operate.

(*The following control signals are optional.*)

(d) DATA TERMINAL READY (*to the modem*). When the terminal sends the ON condition on this connection it causes the modem to be connected to the communication line. The OFF condition causes it to be disconnected, in order to terminate a call or free the line for a different use.

(e) RING INDICATOR (*to the terminal*). A signal on the connection informs the terminal that the modem is receiving a ringing signal from a remote location.

(f) DATA CARRIER DETECTOR (*to the terminal*). A signal on this connection indicates to the terminal that the carrier (the sine wave that carries the signal) is being received. If the carrier is lost because of a fault condition on the line, the terminal will be notified by an OFF condition in this connection.

(g) DATA MODULATION DETECTOR (*to the terminal*). An ON condition on this connection informs the terminal that the signal is being demodulated correctly by the modem. When the quality of demodulation drops below a certain threshold the terminal may take corrective action such as requesting retransmission or requesting that a lower transmission rate be used.

(h) SPEED SELECTOR. There are two speed selector connections, one to the modem and one to the terminal. Using them, the transmission rate may be changed.

4. ***Grounds.***

(a) PROTECTIVE GROUND. Attached to the machine frame and possibly to external grounds.

(b) SIGNAL GROUND. Establishes the common ground reference potential for the circuits.

Other connectors are for optional use at the discretion of the equipment designer.

This RS-232-C Standard and CCITT V. 24 interface apply to serial transmission. Similar standards are used for parallel transmission.

6 CHARACTERS AND CODES

Ever since the earliest swinging telegraph needles and the first waving of semaphore flags it has been necessary to use codes to translate signals transmitted into digits, letters, or words. The Morse code has been learned by Boy Scouts for more than half a century.

Figure 6.1 shows an old-established method of interpreting the ON and OFF pulses of telegraphy into meaningful information. The requirements changed somewhat, in recent years, with the advent of computer data transmission. A proliferation of new codes threatened the tranquil telegraph scene, and, as on all such occasions, a battle for standardization ensued. This chapter discusses the codes now in most common use.

BITS PER CHARACTER

If we transmit n bits such as the ON and OFF pulses of telegraphy we can, in theory, code 2^n different combinations into these pulses. The "characters" that are sent by data transmission often contain 5, 6, 7, or 8 bits. Five bits can give 32 different characters, six bits 64, seven bits 128, and eight bits 256. A seven-bit code is used, for example, to transmit a range of up to 128 characters. It would not always transmit 128 characters, because some of the combinations are reserved for special-purpose *control characters*, which have such functions as indicating the end of a record, making a teleprinter carriage return, and a variety of other operations.

ESCAPE CHARACTERS

Telegraphy has commonly used a five-bit code, as shown in Figure 6.1. This code would restrict the number of characters to 32 if used normally; however, letters shift and figures shift characters are used to extend the range.

When a *figures shift* character is sent, the characters that follow it are uppercase characters until a *letters shift* character is sent. Similarly, the characters following a *letters shift* are letters until a *figures shift* is sent. The *letters shift* and *figures shift* characters must be recognized in either case.

This is an example of what is known as an *escape* mechanism in a code. The "letters shift" and "figures shift" are referred to as *escape* characters. By using escape characters or combinations of characters, the total number of possible characters in a code can be greatly increased. Sometimes the escape character changes the meaning of all the other characters following it until an additional escape character is received, as in the above example. In other codes, the escape character changes the meaning of only the one character that follows it. Thus special characters can be inserted at the wish of the user. Such is the case in the military Fieldata telegraphy code shown in Figure 6.2. The character labeled "Special" here is the escape character, and the character immediately following it has a special alternate meaning.

An important factor in the transmission of computer data is that the computer may have to send *every possible combination* of bits. This situation occurs in the transmission of both data and programs. The Baudot code in Figure 6.1 cannot transmit all bit combinations.

PARITY BITS

Many data transmission codes use an extra bit, called a parity bit, in each character for checking purposes. This parity bit is added so that the total number of 1 bits in the character transmitted will be an odd (or in some codes, even) number. For example, if a character without the parity bit is 0100010, then an extra 1 bit is added to make it 01000101, with an odd number of 1s. On the other hand, if it is 0110010, then a 0 bit is added to again produce an odd number of 1s. The receiving machine detects whether there is an odd number of 1s. If not, it knows that noise or distortion on the line has lost or added a bit. It either notes this error or instigates retransmission of that data. This process is sometimes called *vertical redundancy checking.*

Unfortunately, noise on the communication line sometimes changes more than one bit, thereby lessening the effectiveness of parity checking. If two 0 bits are both changed to 1s, the parity check will not detect the change. Further checking facilities are needed, as discussed in Chapter 8.

Figure 6.1. The Baudot five-bit telegraphy code (CCITT Alphabet No. 2). *Note*: This is commonly referred to as the Baudot code. It was, however, invented by Donald Murray. Baudot's work produced a quite different code structure which resulted in the CCITT Alphabet No. 1. There is little resemblance between the two codes except that they both used five bits per character.

● Denotes positive current

Code signals:

Start	1	2	3	4	5	Stop	Lower case	Upper case: CCITT standard international telegraph alphabet No.2	Upper case: U.S.A. teletype commercial keyboard	Upper case: AT & T fractions keyboard	Upper case: Weather keyboard
	●	●				●	A	–	–	–	↑
	●			●	●	●	B	?	?	5/8	⊕
		●	●	●		●	C	:	:	1/8	◯
	●			●		●	D	Who are you?	$	$	↗
	●					●	E	3	3	3	3
	●		●	●		●	F	Note 1	!	1/4	→
		●		●	●	●	G	Note 1	&	&	↘
			●		●	●	H	Note 1	#		↓
		●	●			●	I	8	8	8	8
	●	●		●		●	J	Bell	Bell	'	↙
	●	●	●	●		●	K	(	(	1/2	←
		●			●	●	L	)	)	3/4	↖
			●	●	●	●	M	.	.	.	.
			●	●		●	N	,	,	7/8	⦶
				●	●	●	O	9	9	9	9
		●	●		●	●	P	0	0	0	∅
	●	●	●		●	●	Q	1	1	1	1
		●		●		●	R	4	4	4	4
	●		●			●	S	,	,	Bell	Bell
					●	●	T	5	5	5	5
	●	●	●			●	U	7	7	7	7
		●	●	●	●	●	V	=	;	3/8	⦶
	●	●			●	●	W	2	2	2	2
	●		●	●	●	●	X	/	/	/	/
	●		●		●	●	Y	6	6	6	6
	●				●	●	Z	+	"	"	+
						●	Blank				–
	●	●	●	●	●	●	Letters shift				↓
	●	●		●	●	●	Figures shift				↑
			●			●	Space				■
				●		●	Carriage return				<
		●				●	Line feed				≡

Note 1: Not allocated internationally; available to each country for internal use.

Standard code	Paper tape code	Data character
8765 4321		
0100 0000	1100 0000	Master space
1100 0001	0100 0001	Upper case
1100 0010	0100 0010	Lower case
0100 0011	1100 0011	Line feed
1100 0100	0100 0100	Carriage return
0100 0101	1100 0101	Space
0100 0110	1100 0110	A
1100 0111	0100 0111	B
1100 1000	0100 1000	C
0100 1001	1100 1001	D
0100 1010	1100 1010	E
1100 1011	0100 1011	F
0100 1100	1100 1100	G
1100 1101	0100 1101	H
1100 1110	0100 1110	I
0100 1111	1100 1111	J
1101 0000	0101 0000	K
0101 0001	1101 0001	L
0101 0010	1101 0010	M
1101 0011	0101 0011	N
0101 0100	1101 0100	O
1101 0101	0101 0101	P
1101 0110	0101 0110	Q
0101 0111	1101 0111	R
0101 1000	1101 1000	S
1101 1001	0101 1001	T
1101 1010	0101 1010	U
0101 1011	1101 1011	V
1101 1100	0101 1100	W
0101 1101	1101 1101	X
0101 1110	1101 1110	Y
1101 1111	0101 1111	Z
1110 0000	1010 0000	)
0110 0001	0010 0001	–
0110 0010	0010 0010	+
1110 0011	1010 0011	<
0110 0100	0010 0100	=
1110 0101	1010 0101	>
1110 0110	1010 0110	–
0110 0111	0010 0111	S
0110 1000	0010 1000	(*)
1110 1001	1010 1001	(
1110 1010	1010 1010	"
0110 1011	0010 1011	:
1110 1100	1010 1100	?
0110 1101	0010 1101	!
0110 1110	0010 1110	.
1110 1111	1010 1111	⊕ (Stop)
0111 0000	0011 0000	0
1111 0001	1011 0001	1
1111 0010	1011 0010	2
0111 0011	0011 0011	3
1111 0100	1011 0100	4
0111 0101	0011 0101	5
0111 0110	0011 0110	6
1111 0111	1011 0111	7
1111 1000	1011 1000	8
0111 1001	0011 1001	9
0111 1010	0011 1010	'
1111 1011	1011 1011	;
0111 1100	0011 1100	/
1111 1101	1011 1101	.
1111 1110	1011 1110	□ Special
0111 1111	0011 1111	Idle
0001 1010	1001 1010	WRU
1011 1111	1111 1111	Delete

Sequence of bits in serial transmission (→ 8765 4321)

Parity bit

Control bit

Data bits

These are control characters because the seventh bit of the standard code is 0 (The sixth and seventh bits of the tape code are the same)

Figure 6.2. Fieldata code used in military data transmission.

The Fieldata code given in Figure 6.2 uses *odd* parity checking as described above for normal transmission. However, it uses even parity checking for its paper tape; in other words, there must be an even number of holes in the tape for each character. This then complies with the convention that a section of tape with all holes punched is valid, but contains no data. To delete incorrect data, one punches *all* the holes of those characters. The reading machine then passes over this section, taking no action.

CONTROL BITS AND CHARACTERS

A variety of control functions are needed in data transmission, some of which were discussed above. It may be necessary, for example, to make a distant printer space or feed a line. In some transmission, it is necessary to indicate end of record, end of text, and so on. Figure 6.1 lists five control characters, at the bottom of the list, and these characters must be recognizable in either figures shift or letters shift. In a similar manner, the more extensive code shown in Figure 6.3 must be able to control distant machines, and here 18 control characters are listed, which again must be recognizable in either shift. The number is often greater than this. Figure 6.6 lists 33 control characters.

The preceding examples have used a special *character* for control. In some codes, a special *bit* is used. This is the case in the Fieldata code of Figure 6.2. Here the first six bits are normally data bits. In the standard code, however, the seventh bit is used for control purposes. If it is a 1, the first six bits *are* data bits. If it is a 0, then the whole character becomes our control character. Thus in Figure 6.2 all the characters except the last two in the list have a 1 as the seventh bit of the standard code (not the paper-tape code). They are therefore data characters. The last two are control characters because their seventh bit is 0 (otherwise they would be "u" and "Idle," respectively). In the paper-tape version of this code the convention is somewhat different. Here the characters are control characters if and only if the sixth and seventh bits are the same. Thus all the paper-tape code characters in Figure 6.2, except the last two, have a different sixth and seventh bit. In the last two characters, these are the same.

BINARY-CODED DECIMAL

In thc early days of data proccssing, the main medium for storing data was punched cards, and a number of punched-card codes became standardized

When the card is transmitted, it is desirable to make the transmitting machine as inexpensive as possible, and so a code has been used that is close to that on the card. Codes referred to as binary-coded decimal are used to represent in six, or sometimes seven, bits the data on the card.

The Hollerith card, a widely accepted standard, uses, in effect, 12 bits

DATA CHARACTERS:

Normal shift

Character:	S	B	A	8	4	2	1	Parity
1	0	0	0	0	0	0	1	0
2	0	0	0	0	0	1	0	0
3	0	0	0	0	0	1	1	1
4	0	0	0	0	1	0	0	0
5	0	0	0	0	1	0	1	1
6	0	0	0	0	1	1	0	1
7	0	0	0	0	1	1	1	0
8	0	0	0	1	0	0	0	0
9	0	0	0	1	0	0	1	1
0	0	0	0	1	0	1	0	1
a	0	1	1	0	0	0	1	0
b	0	1	1	0	0	1	0	0
c	0	1	1	0	0	1	1	1
d	0	1	1	0	1	0	0	0
e	0	1	1	0	1	0	1	1
f	0	1	1	0	1	1	0	1
g	0	1	1	0	1	1	1	0
h	0	1	1	1	0	0	0	0
i	0	1	1	1	0	0	1	1
j	0	1	0	0	0	0	1	1
k	0	1	0	0	0	1	0	1
l	0	1	0	0	0	1	1	0
m	0	1	0	0	1	0	0	1
n	0	1	0	0	1	0	1	0
o	0	1	0	0	1	1	0	0
p	0	1	0	0	1	1	1	1
q	0	1	0	1	0	0	0	1
r	0	1	0	1	0	0	1	0
s	0	0	1	0	0	1	0	1
t	0	0	1	0	0	1	1	0
u	0	0	1	0	1	0	0	1
v	0	0	1	0	1	0	1	0
w	0	0	1	0	1	1	0	0
x	0	0	1	0	1	1	1	1
y	0	0	1	1	0	0	0	1
z	0	0	1	1	0	0	1	0
.	0	1	1	1	0	1	1	0
$	0	1	0	1	0	1	1	1
,	0	0	1	1	0	1	1	1
/	0	0	1	0	0	0	1	1
'	0	0	0	1	0	1	1	0
&	0	1	1	0	0	0	0	1
—	0	1	0	0	0	0	0	0
@	0	0	1	0	0	0	0	0

Upper shift

Character:	S	B	A	8	4	2	1	Parity
=	1	0	0	0	0	0	1	1
c	1	0	0	0	0	1	0	1
;	1	0	0	0	0	1	1	0
:	1	0	0	0	1	0	0	1
°C	1	0	0	0	1	0	1	0
'	1	0	0	0	1	1	0	0
—	1	0	0	0	1	1	1	1
+	1	0	0	1	0	0	0	1
(	1	0	0	1	0	0	1	0
)	1	0	0	1	0	1	0	0
A	1	1	1	0	0	0	1	1
B	1	1	1	0	0	1	0	1
C	1	1	1	0	0	1	1	0
D	1	1	1	0	1	0	0	1
E	1	1	1	0	1	0	1	0
F	1	1	1	0	1	1	0	0
G	1	1	1	0	1	1	1	1
H	1	1	1	1	0	0	0	1
I	1	1	1	1	0	0	1	0
J	1	1	0	0	0	0	1	0
K	1	1	0	0	0	1	0	0
L	1	1	0	0	0	1	1	1
M	1	1	0	0	1	0	0	0
N	1	1	0	0	1	0	1	1
O	1	1	0	0	1	1	0	1
P	1	1	0	0	1	1	1	0
Q	1	1	0	1	0	0	0	0
R	1	1	0	1	0	0	1	1
S	1	0	1	0	0	1	0	0
T	1	0	1	0	0	1	1	1
U	1	0	1	0	1	0	0	0
V	1	0	1	0	1	0	1	1
W	1	0	1	0	1	1	0	1
X	1	0	1	0	1	1	1	0
Y	1	0	1	1	0	0	0	0
Z	1	0	1	1	0	0	1	1
.	1	1	1	1	0	1	1	1
!	1	1	0	1	0	1	1	0
,	1	0	1	1	0	1	1	0
?	1	0	1	0	0	0	1	0
±	1	0	0	1	0	1	1	1
+	1	1	1	0	0	0	0	0
—	1	1	0	0	0	0	0	1
*	1	0	1	0	0	0	0	1

CONTROL CHARACTERS

(Either shift)
(Either setting of S bit)

Backspace			1	0	1	1	1	0	
End of transfer			0	0	1	1	1	1	
Delete			1	1	1	1	1	1	
Down-shift			1	1	1	1	1	0	
Carriage return			1	0	1	1	0	1	
Prefix			0	1	1	1	1	1	
Idle			1	0	1	1	1	1	
Reader stop			0	0	1	1	0	1	
Space			0	0	0	0	0	0	
End of block			0	1	1	1	1	0	
Up-shift			0	0	1	1	1	0	
Line feed			0	1	1	1	0	1	
Tab			1	1	1	1	0	1	
Restore			1	0	1	1	0	0	
Bypass			0	1	1	1	0	0	
End of heading			0	0	1	0	1	1	
Punch on			0	0	1	1	0	0	
Punch off			1	1	1	1	0	0	

Figure 6.3. BCD coding used on various IBM terminals, using seven data bits and a parity check bit.

to code one character that could be coded with six bits. The coding is therefore condensed to produce the binary-coded decimal (BCD) form.

This may be seen by comparing the normal shift of the BCD code in Figure 6.3 with the coding on the Hollerith card. The first bit in Figure 6.3 decides which shift the transmission is in. The second shift gives an extra set of characters—fewer than 64, because some are used for control characters. The upper-shift characters of Figure 6.3 therefore have the same B, A, 8, 4, 2, 1 bits as the normal shift. Only the S bit (and hence the parity bit) is different.

It should be noted that one computer manufacturer's BCD code can be entirely different from another's. The Sperry Rand code is entirely different from the IBM code.

EIGHT-BIT BCD Figure 6.4 shows an extension of the BCD code to use eight bits. This has plenty of space to spare for special characters. It is used to transmit the eight-bit bytes of computers like the IBM 360 or 370, or to operate a printing device with up to 256 graphic characters.

Bit positions 1,2,3,4

Bit positions 5,6,7,8	0000	0001	0010	0011	0100	0101	0110	0111	1000	1001	1010	1011	1100	1101	1110	1111
0000	NUL				Blank	B	–						>	<	t	0
0001							/		a	j			A	J		1
0010									b	k	s		B	K	S	2
0011									c	l	t		C	L	T	3
0100	PF	RES	BYP	PN					d	m	u		D	M	U	4
0101	HT	NL	LF	RS					e	n	v		E	N	V	5
0110	LC	BS	EOB	UC					f	o	w		F	O	W	6
0111	DEL	IDL	PRE	EOT					g	p	x		G	P	X	7
1000									h	q	y		H	Q	Y	8
1001							/	″	i	r	z		I	R	Z	9
1010					?	!		:								
1011					.	S	,	#								
1100					←	*	%	@								
1101					(	)	˅	'								
1110					+	;	–	=								
1111					‡	¢	±	✓								

Figure 6.4 An extended BCD code with eight bits and plenty of space to spare for special characters. Letter A, for example, is 11000001. This code is used to transmit the eight-bit bytes of computers using this code, or to operate a printing device with up to 256 graphic characters.

N-OUT-OF-M CODES

It was mentioned earlier that a parity bit used for error detection, as in Figures 6.2 and 6.3, is not as effective as one would like because transmission noise often changes more than one bit in a character. This noise can be counteracted in two ways: first, by using a separate error-checking pattern at the end of a record or message, as will be discussed in Chapter 8, and, second, by coding self-checking characters in a more effective way than those above.

One way of coding characters to detect the loss or addition of a small group of bits is to use an *N*-out-of-*M* code. A number, *M*, of bits are used to transmit each character. *N* of these *M* bits will always be 1s [and $(M - N)$ will always be 0s] if the character is received correctly. A common example is a 4-out-of-8 code. Here eight bit characters are coded in such a way that four bits are always 1s and four bits are always 0s. Then if any group of bits is lost from or added to a character, the error-detection circuit will spot this (unless the character is completely wiped out). It is, of course, possible that one noise pulse could add a bit and another delete a bit.

The right-hand side of Figure 6.5 illustrates the coding of characters using a 4-out-of-8 code. Seventy combinations are possible. This number is far fewer than the 128 combinations of an eight-bit code with parity, and that is the price one pays for the added safety. Error-checking patterns can be devised that give both a lower level of redundancy and a higher level of protection than *N*-out-of-*M* codes, but they require more logic for validity checking.

Suppose that a data-processing machine that employs seven-bit BCD coding transmits using the 4-out-of-8 codes. This situation is illustrated in Figure 6.5. The 4-out-of-8 code has six more characters than the BCD code. These are used for transmission control characters that govern the sending of data but that do not enter into the data processing.

The machines at both ends must translate from one code to the other. This step has conventionally been done by using a specially wired core plane so that characters read into it in one code can be read out in the other. Today it might be done in LSI hardware logic. In computers, the code conversion is sometimes done under program control, possibly by using special instructions wired or microprogrammed into the machine.

AMERICAN STANDARDS ASSOCIATION CODE

A profusion of different data transmission codes has arisen. There are many others beside those discussed in this chapter. Individual machine designers often have their own reason for employing yet another code. A further transmission code, for example, will appear in the next chapter.

As the usage of data transmission grows, and particularly as more and

Character	BCD Code							4-of-8 Code							
	C	B	A	8	4	2	1	1	2	4	8	R	O	X	N
64 DATA CHARACTERS															
Space	C								2	4	8		O		
0	C			8		2			2		8	R			N
1							1	1					O	X	N
2						2			2				O	X	N
3	C					2	1	1	2			R			N
4					4					4			O	X	N
5	C				4		1	1		4		R			N
6	C				4	2			2	4		R			N
7					4	2	1	1	2	4		R			
8				8							8		O	X	N
9	C			8			1	1			8	R			N
A		B	A				1	1				R	O	X	
B		B	A			2			2			R	O	X	
C	C	B	A			2	1	1	2				O	X	
D		B	A		4					4		R	O	X	
E	C	B	A		4		1	1		4			O	X	
F	C	B	A		4	2			2	4			O	X	
G		B	A		4	2	1	1	2	4					N
H		B	A	8							8	R	O	X	
I	C	B	A	8			1	1			8		O	X	
J	C	B					1	1				R		X	N
K	C	B				2			2			R		X	N
L		B				2	1	1	2					X	N
M	C	B			4					4		R		X	N
N		B			4		1	1		4				X	N
O		B			4	2			2	4				X	N
P	C	B			4	2	1	1	2	4				X	
Q	C	B		8							8	R		X	N
R		B		8			1	1			8			X	N
S	C		A			2			2			R	O		N
T			A			2	1	1	2				O		N
U	C		A		4					4		R	O		N
V			A		4		1	1		4			O		N
W			A		4	2			2	4			O		N
X	C		A		4	2	1	1	2	4			O		
Y	C		A	8							8	R	O		N
Z			A	8			1	1			8		O		N
/	C		A				1	1				R	O		N
#				8		2	1	1	2		8	R			
.		B	A	8		2	1	1	2		8				N
$	C	B		8		2	1	1	2		8			X	
'	C		A	8		2	1	1	2		8		O		
@	C			8	4					4	8	R			N
□	C	B	A	8	4					4	8		O	X	
•		B		8	4					4	8			X	N
&	C	B	A						2	4	8				N
%			A	8	4					4	8		O		N
—		B							2	4	8			X	
+0	C	B	A	8		2			2		8		O	X	
-0		B		8		2			2		8			X	N
RM			A	8		2			2		8		O		N
GM	C	B	A	8	4	2	1	1		4	8				N
Delta		B		8	4	2	1	1		4	8			X	
SM			A	8	4	2	1	1		4	8		O		
WS	C		A	8	4		1		2		8	R		X	
(		B	A	8	4		1		2	4		R		X	
?		B	A	8	4	2			2	4		R	O		
)	C	B		8	4		1			4	8	R		X	
;	C	B		8	4	2				4	8	R	O		
,	C		A	8	4	2			2		8	R	O		
"				8	4	2			2	4	8	R			
:				8	4		1	1	2	4	8				
TM				8	4	2	1	1		4	8	R			
6 Control Characters:															
Idle								1			8	R	O		
Inquiry/Error								1			8	R		X	
Transmit Leader								1		4		R	O		
Control Leader								1		4		R		X	
SOR1/ACK1 (Start of Record/Ack)								1	2			R		X	
SOR2/ACK2 (Start of Record/Ack)								1	2			R	O		
Dual Characters:															
TEL (Telephone										4	8	R		X	
EOF (End of File)									2		8	R		X	

These cannot be translated into the 6-bit BCD code

Note:

The 4 out of 8 code permits 70 combinations

The 6-bit BCD code permits 64 combinations

Figure 6.5. An example of coding in a 4-out-of-8 transmission code to transmit data coded in a six-bit BCD code.

more machines are used to dial up other machines on the public network, so the need for standardization of transmission codes becomes great. The American Standards Association, CCITT, and other national bodies have given much thought to this problem. The requirements of different users conflict considerably; consequently, there are difficulties in agreeing on standards.

The American Standards Association, however, has standardized the U.S. ASCII shown in Figure 6.6. (ASCII stands for American Standard Code for Information Interchange.) This is a seven-bit code that permits double shift printing and has enough control characters for most purposes. An eighth bit can be used if desired as a parity check.

This code has come into wide acceptance in telegraphy and data transmission in the United States. Most of the major computer manufacturers are using it in their equipment, both in the United States and other countries.

The code in Figure 6.6 is a seven bit-code. It would be of value to have a six-bit and eight-bit standard code as well. The American Standards Association is working on this, but no such codes are accepted at the time of writing. Figure 6.7 shows a six-bit code used by IBM that contains a subset of the characters in the seven-bit ASCII code. IBM's "binary synchronous" range of transmission machines use the six-bit, seven-bit, and eight-bit codes of Figures 6.7, 6.6, and 6.4.

	Bit positions 5,6,7							
Bit positions 1,2,3,4	000	100	010	110	001	101	011	111
0000	NUL	DLE	SPACE	0	@	P	`	p
1000	SOH	DC1	!	1	A	Q	a	q
0100	STX	DC2	"	2	B	R	b	r
1100	ETX	DC3	#	3	C	S	c	s
0010	EOT	DC4	$	4	D	T	d	t
1010	ENQ	NAK	%	5	E	U	e	u
0110	ACK	SYN	&	6	F	V	f	v
1110	BEL	ETB	'	7	G	W	g	w
0001	BS	CAN	(	8	H	X	h	x
1001	HT	EM	)	9	I	Y	i	y
0101	LF	SUB	*	:	J	Z	j	z
1101	VT	ESC	+	;	K	[	k	{
0011	FF	FS	⌐	<.	L	\	l	|
1011	CR	GS	–	=	M	]	m	}
0111	SO	RS	. ·	>	N	∧	n	~
1111	SI	US	/	?	O	–	o	DEL

Figure 6.6. American Standard Code for Information Interchange (ASCII).

NUL (Null): No character. Used for filling in time or filling space on tape when there is no data.

SOH (Start of Heading): Used to indicate the start of a heading which may contain address or routing information.

STX (Start of Text): Used to indicate the start of the text and so also indicates the end of the heading.

ETX (End of Text): Used to terminate the text which was started with STX.

EOT (End of Transmission): Indicates the end of a transmission, which may have included one or more "texts" with their headings.

ENQ (Enquiry): A request for a response from a remote station. It may be used as a "WHO ARE YOU?" request for a station to identify itself.

ACK (Acknowledge): A character transmitted by a receiving device as an affirmation response to a sender. It is used as a positive response to polling messages.

BEL (Bell): Used when there is need to call human attention. It may control alarm or attention devices.

BS (Backspace): Indicates movement of the printing mechanism or display cursor backwards in one position.

HT (Horizontal Tab): Indicates movement of the printing mechanism or display cursor forward to the next preassigned "tab" or stopping position.

LF (Line Feed): Indicates movement of the printing mechanism or display cursor to the start of the next line.

VT (Vertical Tab): Indicates movement of the printing mechanism or display cursor to the next of a series of preassigned printing lines.

FF (Form Feed): Indicates movement of the printing mechanism or display cursor to the starting position of the next page, form, or screen.

CR (Carriage Return): Indicates movement of the printing mechanism or display cursor to the starting position of the same line.

SO (Shift Out): Indicates that the code combinations which follow shall be interpreted as *outside* of the standard character set until a SHIFT IN character is reached.

SI (Shift In): Indicates that the code combinations which follow shall be interpreted according to the standard character set.

DLE (Data Link Escape): A character which shall change the meaning of one or more contiguously following characters. It can provide supplementary controls, or permits the sending of data characters having any bit combination.

DC1, DC2, DC3 and DC4 (Device Controls): Characters for the control of ancillary devices or special terminal features.

NAK (Negative Acknowledgment): A character transmitted by a receiving device as a negative response to a sender. It is used as a negative response to polling messages.

SYN (Synchronous/Idle): Used as a synchronous transmission system to achieve synchronization. When no data is being sent a synchronous transmission system may send SYN characters continuously.

ETB (End of Transmission Block): Indicates the end of a block of data for communication purposes. It is used for blocking data where the block structure is not necessarily related to the processing format.

CAN (Cancel): Indicates that the data which precedes it in a message or block should be disregarded (usually because an error has been detected).

EM (End of Medium): Indicates the physical end of a card, tape or other medium, or the end of the required or used portion of the medium.

SUB (Substitute): Substituted for a character that is found to be erroneous or invalid.

ESC (Escape): A character intended to provide code extension in that it gives a specified number of contiguously following characters an alternate meaning.

FS (File Separator):
GS (Group Separator):
RS (Record Separator):
US (United Separator):
Information separators to be used in an optional manner except that their hierarchy shall be FS (the most inclusive) to US (the least inclusive).

SP (Space): A nonprinting character used to separate words, or to move the the printing mechanism or display cursor forward by one position.

DEL (Delete): Used to obliterate unwanted characters (for example, on paper tape by punching a hole in *every* bit position).

Figure 6.6.cont.

Bit positions 3, 4, 5, 6	Bit positions 1, 2: 00	01	10	11
0000	SOH	&	–	0
0001	A	J	/	1
0010	B	K	S	2
0011	C	L	T	3
0100	D	M	U	4
0101	E	N	V	5
0110	F	O	W	6
0111	G	P	X	7
1000	H	Q	Y	8
1001	I	R	Z	9
1010	STX	SPACE	ESC	SYN
1011		$	,	
1100	<	*	%	@
1101	BEL	US	ENQ	NAK
1110	SUB	EOT	ETX	EM
1111	ETB	DLE	HT	DEL

Figure 6.7. A six-bit code used by IBM containing a subset of the characters in the seven-bit American standard code, ASCII. The control characters have the same meanings as those listed in Figure 6.6.

TRANSPARENT CODES

All the foregoing codes contain control characters. The transmission cannot take place without some of these. However, as we commented earlier, it is often desirable to transmit *all* of the six-, seven-, or eight-bit combinations that a computer or its peripheral devices can store.

One way to resolve this conflict is by using a *pair* of characters for the control character, instead of one. For example, the DLE escape character (Data Link Escape) of the ASCII and similar codes may precede any control character and this tells the receiving machine that the control character has its control meaning. The DLE character is regarded as not being part of the data. To transmit a DLE character and have the receiving machine accept it, it must itself be preceded by a DLE character.

This type of transmission is sometimes referred to as a *transparent* code or transmission in *transparent* text mode.

Some machines can switch backward and forward between transparent and normal text. Sequences of characters are needed to do so, such as the following:

DLE STX: Initiate transparent text mode.

DLE ETB: Terminate transparent transmission.

DLE ITB: Terminate transparent text mode but continue transmission in normal mode.

Another approach is to use unique bit patterns for delineating the start and end of a transmission block, and then any combination of bits can be placed within the block. Suppose, for example, that six consecutive 1 bits are made a unique pattern. This pattern could be prevented from being present in the data by using a logic circuit on transmission that looks constantly for five consecutive 1 bits, and if it finds them, it automatically inserts a 0 bit. The receiving device has an equivalent logic circuit that looks constantly for the bit combination 111110. When it finds it, it deletes the 0 bit. In this way, any combination of bits can be sent without ever containing six consecutive 1 bits. A pattern such 01111110 can then be used for indicating the start and end of a block, and inside the block there can be any bit pattern.

MINIMAL ENCODING

Efficiency of encoding is important on some data transmission networks, for it can enable an expensive communication line to serve the maximum number of terminals. If any criticism can be leveled at the widely accepted ASCII code, it is that seven bits are unnecessary for many applications; six bits would be enough. This is often the case in commercial applications, in which only an alphanumeric character set is needed. Some commercial data-processing systems using long lines have therefore chosen a code such as that in Figure 6.7. In tightly engineered systems, no parity bits were used on the characters because (as we will discuss later) block error-detection encoding is more efficient than character error detection.

The encoding for commercial applications could be made still tighter by using a five-bit character set that may have three shifts rather than two like the Baudot code. In fact, a third shift could be added to the Baudot code easily, which could have three advantages [1]:

1. Efficient encoding of alphanumeric information.
2. Complete transparency, permitting the transmission of any combination of bits.
3. Terminals could be made inexpensively compatible with Baudot code machines which are used on the vast international Telex network.

7 MODES OF TRANSMISSION

There are a variety of ways in which digital data can be transmitted over a given transmission line; in other words, a variety of methods of organizing the signals sent so that they convey the information in question. For each different transmission mode, there are families of input-output machines, computer transmission line adapters, and so on, built to operate in this manner.

FULL DUPLEX VERSUS HALF DUPLEX

Over a given physical line, the terminal equipment may be designed so that it can either transmit in both directions at once—*full-duplex* transmission—or else so that it can transmit in either direction but not both at the same time—*half-duplex*.

A terminal, or a computer transmission control unit, works in somewhat different fashions, depending on which of these possibilities is used. Where full-duplex transmission is employed, it may be used either to send data streams in both directions at the same time or to send data in one direction and control signals in the other. The control signals govern the flow of data and are used for error control. Data at the transmitting end are held until the receiving end indicates that the data have been received correctly. If the data are not received correctly, the control signal indicates this, and the data are retransmitted. Control signals ensure that no two terminals transmit at once on a line with many terminals, and organize the sequence of transmission.

Simultaneous transmission in two directions can be obtained on a two-wire line by using two separate frequency bands. One is used for transmission in one direction and the other for the opposite direction. By keeping the signals strictly separated in frequency, they can be prevented from interfering with each other.

The two bands may not be of the same bandwidth. A much larger channel capacity is needed for sending data than for sending the return signals that control the flow of data. If, therefore, data are to be sent in one direction only, the majority of the line bandwidth can be used for data.

One modem (not available commercially at the time of writing) permits transmission of data at 3600 bits per second in one direction and provides a simultaneous return path for control signals at 150 bits per second. For many systems this is ideal, and we may see it come into more common use.

Many data-processing situations are not able to take advantage of the facility to transmit streams of data in both directions at the same time. Consequently, where full-duplex transmission is used, it is often with data traveling in one direction only, the other direction being used for control signals.

PARALLEL VERSUS SERIAL TRANSMISSION

Digital data can be sent over communication lines in either a serial mode or a parallel mode. The stream of data is often divided into characters (as with the characters printed by a teleprinter). The characters are composed of bits. This stream may be sent either serial by character and serial by bit, or serial by character and parallel by bit.

Let us suppose that the characters are composed of six bits each. The serial-by-character, parallel-by-bit system must then transmit six bits at once. This is done on some terminals by using six separate communication paths as shown in Figure 7.1. In practice, there would usually be a seventh for control purposes. This is not likely to be done on long-distance, low-speed lines, for it would be more expensive than other means. Several low-speed lines are more expensive than one higher-speed line. Multiple wires are often used, however, within a customer's premises.

Parallel-wire transmission does have the advantage that it can lower the terminal cost. No circuitry is needed in the terminals for deciding which bits are which in a character. Parallel-wire transmission is, therefore, commonly used over short distances where the wires are laid down by the user. For data collection terminals in a factory, for example, which connect to a computer or other machine within that factory, bunches of wires are often installed to connect the machines in a parallel fashion. The terms *in-plant* and *out-plant* system are used. The former means that the communication links are within one plant or localized area, and it implies that common carrier lines are not used.

"Out-plant" implies that common carrier (telecommunication company) lines are used. The economics of what line connections are employed changes when the lines are laid down by the user, rather than leased.

Some machines are designed for parallel-by-bit transmission, but separate parallel wires are not used to connect them. Instead the bits travel simultaneously, using different frequency bands on the same wire. One physical channel is split up into several effective channels, each operating on a different frequency band.

MULTITONE TRANSMISSION

Another form of parallel transmission uses tones such as those generated by a pushbutton telephone. A Bell System Touchtone telephone keyboard can transmit eight audible frequencies: 697, 770, 852, 941, 1209, 1336, 1477, and 1633 hertz. The pressing of any one key produces a discordant combination of two of these frequencies, one from the first four and one from the second. The Bell System 400 Series data sets use the same eight frequencies plus others, and to these a data transmission device operating at 10 characters per second or less may be connected. This is illustrated in Figure 7.2 The IBM 1001 shown in Figure 7.3 operates in this manner. A code is used in which two frequencies out of the eight possible are transmitted at any one instant. This gives 16 possible combinations that can be transmitted. It gives some measure of transmission error detection in that a fault causing only one or more than two, frequences to be received will be noted as an error. This is not comprehensive error detection as in a 4-out-of-8 code.

This code does not have enough combinations for alphabetic transmission. When cards with alphanumeric punches are to be transmitted on the IBM 1001, three more frequencies are needed, one for each alphabetic "zone" punch. The coding remains the same as we have discussed, but now an extra (third) frequency can be sent.

SYNCHRONOUS VERSUS ASYNCHRONOUS TRANSMISSION

Data transmission can be either *synchronous* or *asynchronous*. Asynchronous transmission is often referred to as *start-stop*. With synchronous transmission, characters are sent in a continuous stream. A block of perhaps 100 characters or more may be sent at one time, and for the duration of that block the receiving terminal must be exactly in phase with the transmitting terminal. With asynchronous transmission, one character is sent at a time. The character is initialized by a START signal, shown in Figure 7.4 as a "0" condition on the line, and terminated by a STOP signal, here a "1" condition on the line. The pulses between these two give the

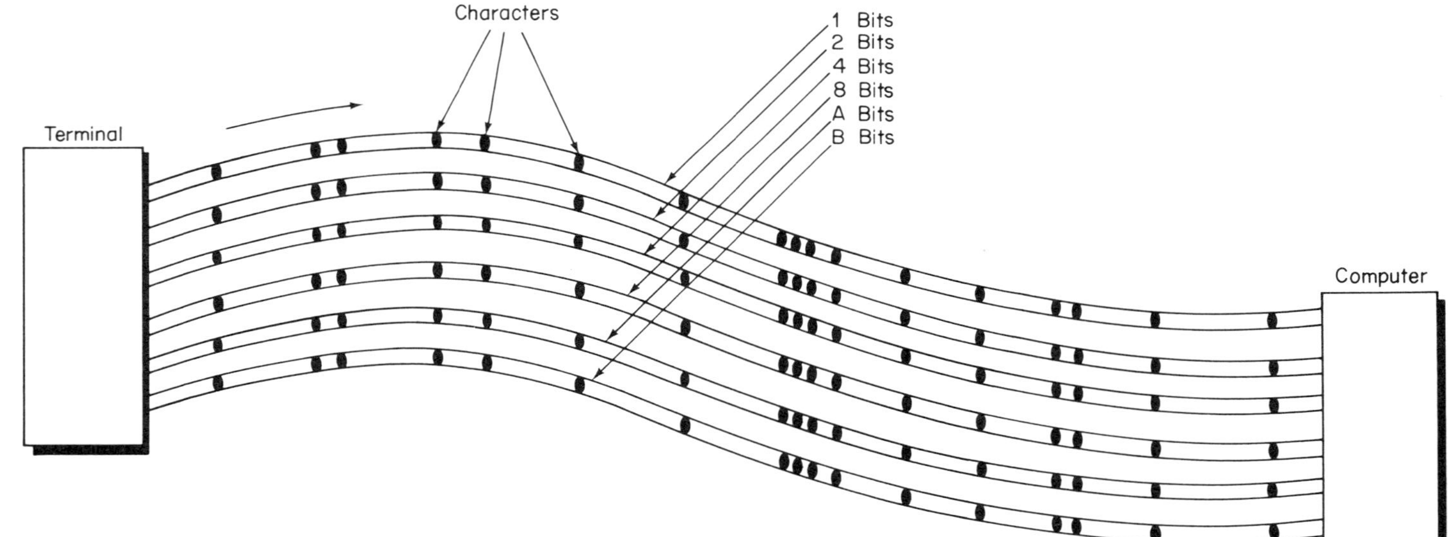

Figure 7.1. Parallel transmission. The different bits in a character each have their own wire or channel. In this way the receiving machine can sort them out easily. The separate paths may not be separate wires but different frequency bands traveling over the same wire.

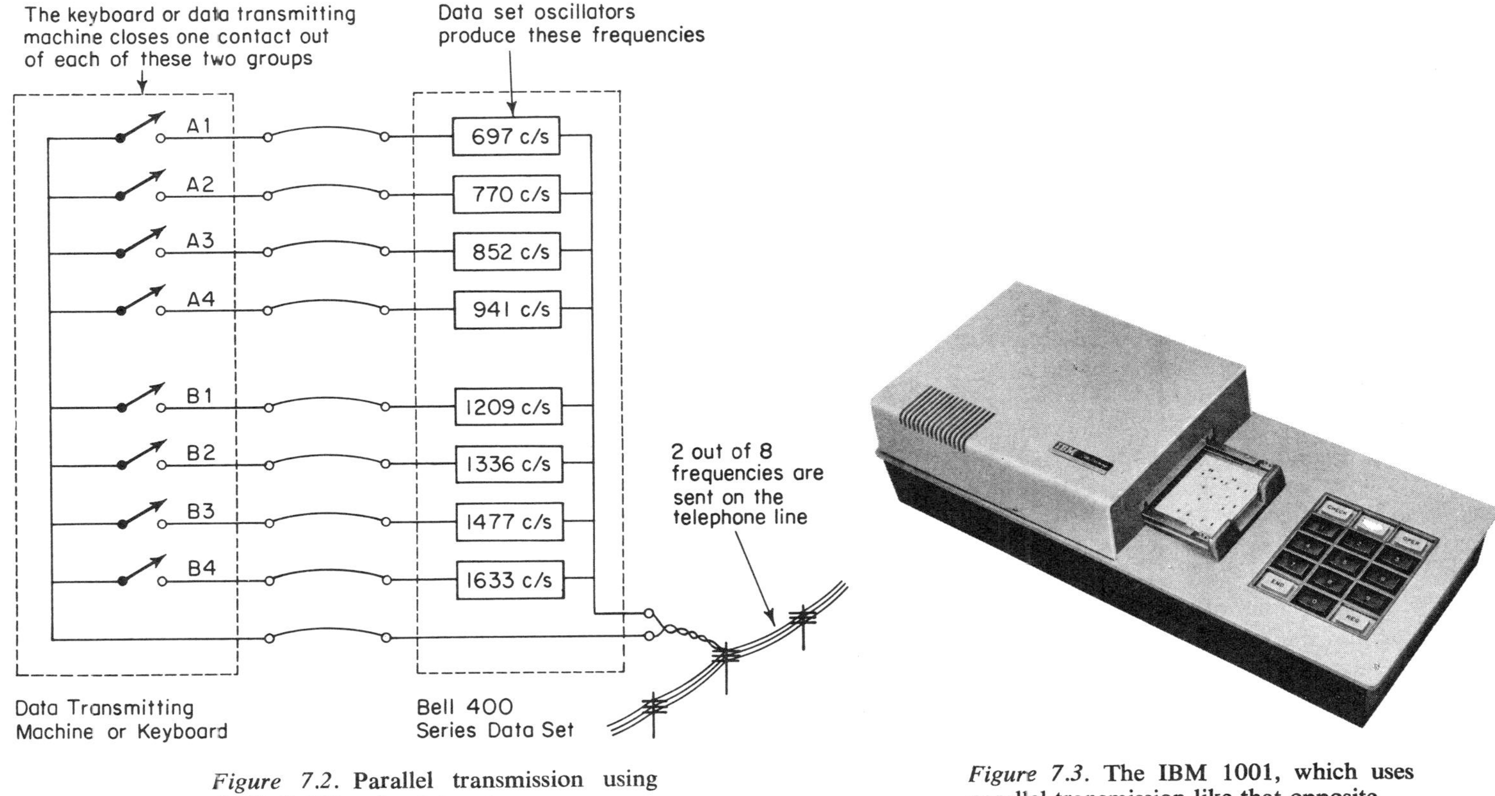

Figure 7.2. Parallel transmission using the Bell 400 series data sets.

Figure 7.3. The IBM 1001, which uses parallel transmission like that opposite.

bits of which the character is composed. Between characters the line is in a "1" condition. As a START bit switches it to "0," the receiving machine starts sampling the bits.

ASYNCHRONOUS (START-STOP) TRANSMISSION

Figure 7.4 shows the form in which a character is sent with start-stop transmission. The two most common types of character coding are illustrated, Baudot code (Figure 6.1) and ASCII code (Figure 6.6). Most non-American teleprinters transmit Baudot characters with five data bits plus the START and STOP elements, as shown at the top of the inset in Figure 7.4. Most American teleprinters designed in recent years transmit ASCII characters, shown at the bottom of Figure 7.4, with eight data bits (of which one is often unused) plus the START and STOP elements.

Start-stop transmission is usually used on keyboard devices which do not have a buffer and on which the operator sends characters along the line at more or less random intervals as she happens to press the keys. The START pulse initiates the sampling, and thus there can be an indeterminate interval between the characters. Characters are transmitted when the operator's finger presses the keys. If the operator pauses for several seconds between one keystroke and the next, the line will remain in the "1" condition for this period of time.

Start-stop machines are generally less expensive to produce than synchronous machines, and for this reason many machines that transmit card-to-card, or paper-tape-to-printer, card-to-computer, and so on, are also start-stop, although the character stream does not have the pauses between characters a keyboard transmission has. Figure 7.5 shows some typical start-stop machines.

The receiving machine has, in essence, a clocking device that starts when the START element is detected and operates for as many bits as there are in a character. With this, the receiving machine can distinguish which bit is which. The STOP element is longer than the data bits in case the receiver clock was not operating at quite the same speed as the transmitter.

When this start-stop transmission is used, there can be an indeterminate period between one character and the next. When one character ends, the receiving device waits idly for the start of the next character. The transmitter and the receiver are then started together, and they remain in phase while the character is sent. The receiver thus is able to attach the correct meaning to each bit it receives.

When an automatic machine such as a paper-tape reader is sending start–stop signals, the length of the STOP condition is governed by the sending machine. It is short, always 1.42 (1.5 or 2) times the other bits, so as to obtain the maximum transmission rate. When a typist uses the keyboard of a start-stop machine, on the other hand, the duration between her keystrokes varies. The transmission occurs when she presses each key, so the stop condition varies in length considerably. When a scientist uses a teleprinter on a time-sharing system, he may be doing work that involves a great amount of thinking. Between one character and another there may occasionally be a very long pause while he thinks or makes notes or drinks his coffee. The STOP "bit" will last for the duration of this pause. The speeds and timing of start-stop machines have become standardized. This means that a variety of different telegraphy machines can communicate with one another. A computer can send data to a telegraph machine by sending pulses with the same timing.

Teletype speeds in common use are listed in Table 7.1. Such speeds are often quoted in "words per minute." An average teletype word is considered to be five characters long. As there is a space character between words, there are, then, six characters per word; and x words per minute $= 10x$ characters per second.

Table 7.1

Speed in Bauds (bits per second)	Bit Time (milliseconds)	Number of Bit Times per Character	STOP element Duration (bit times)	Duration (milliseconds)	Characters per Second	Words per Minute (nominal)
45.5	21.98	7.42	1.42	21.97	6.13	60
50	20	7.42	1.42	20	6.74	66
50	20	7.50	1.50	20	6.67	66
74.2	13.48	7.42	1.42	13.48	10	100
75	13.33	7.50	1.50	13.33	10	100
75	13.33	10	1.00	13.33	7.5	75
75	13.33	11	2.00	13.33	6.82	68
150	6.67	10	1.00	6.67	15	150

A considerable amount of distortion can occur in the length and positioning of the start-stop pulses without the receiver's misinterpreting them. The bit pulse can become shortened or lengthened, or late or early, by an amount somewhat less than half the bit width.

SYNCHRONOUS TRANSMISSION

When machines transmit to each other continuously, with regular timing, *synchronous* transmission can give the most efficient line utilization. Here the bits of one character are followed immediately by those of the next. Be-

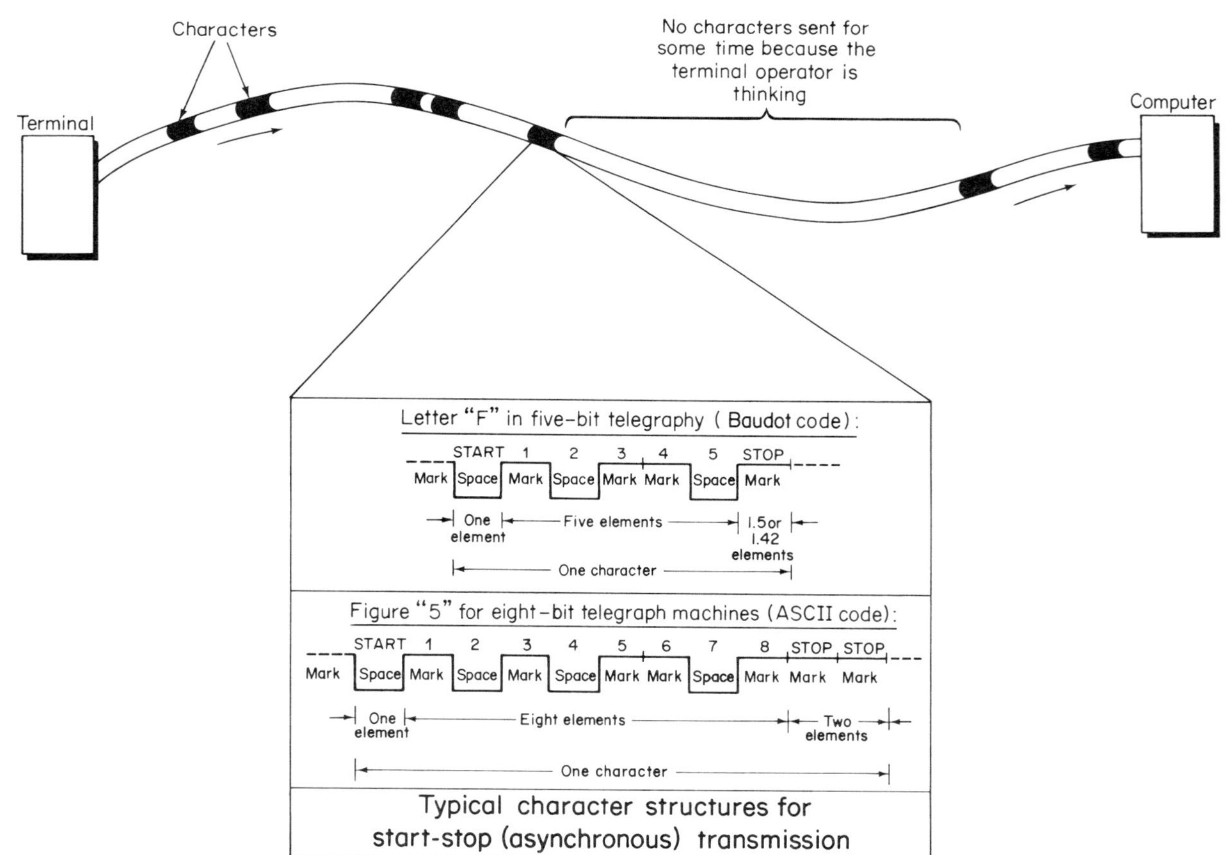

Figure 7.4. Asynchronous (start–stop) transmission.

A typical Telex installation. The user has a paper tape reader, punch, a keyboard, and printer, and may dial other Telex subscribers throughout the world. The machines use Baudot (Murray) code at 50 bits per second (66 words per minute). *Courtesy G.P.O., London*

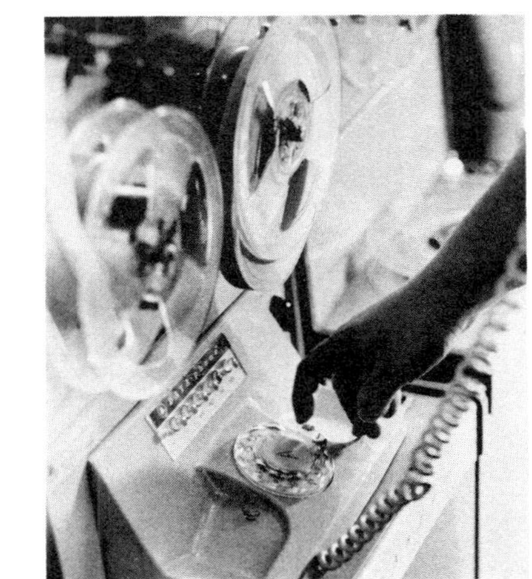

A teletype DATASPEED paper-tape transmitter. This transmits over dial-up or private voice lines at 1050 words per minute, to DATASPEED paper-tape receivers or printers. *Photograph courtesy A.T. & T.*

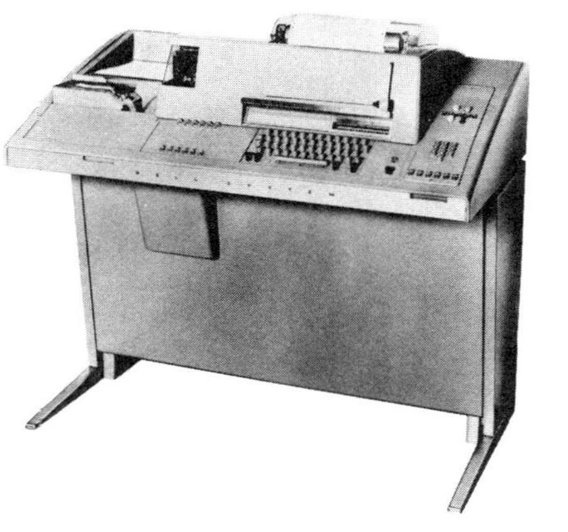

A typical TWX installation, similar to the Telex installation above, but operating in North America (not worldwide like Telex) at speeds up to 150 bits per second. The U.S. ASCII code is used. This Teletype 35 Automatic Send Receive machine is also used on leased lines. *Photograph courtesy A.T. & T.*

***Figure* 7.5. Machines that use start-stop transmission.**

tween characters, there are no START or STOP bits and no pauses (Figure 7.6). The stream of characters of this type is divided into blocks. All the bits in the block are transmitted at equal time intervals. The transmitting and receiving machines must be exactly in synchronization for the duration of the block, so that if the receiving machine knows which is the first bit it will be able to tell which are the bits of each character.

Devices using synchronous transmission have a wide variety of block lengths. The block size may vary from a few characters to many hundreds of characters. Often it relates to the physical nature of the data medium. For example, in the transmission of punched cards it is convenient to use 80 characters as the maximum block length, for there are that many characters per card. Similarly, the length of print lines, the size of buffers, the number of characters in records, or some such system consideration may determine the block size. Some time is taken up between the transmission of one block and the next, so the larger the block length, in general, the faster the overall transmission. One the other hand, the larger the block the higher the probability that it contains an error and will have to be retransmitted. A compromise between these two factors must be evaluated. Figure 7.7 shows some typical machines that use synchronous transmission, and Figure 7.6 shows a typical block of synchronous transmission.

With asynchronous transmission, the unit of transmission is normally the character. The operator of a teletype machine presses a key on her keyboard and one character is sent, complete with its START and STOP bits. It is independent in time of any other character. With synchronous transmission, the characters are stored until a complete block is ready to be sent. The block is sent from a buffer at the maximum speed of the line and its modems. There are no gaps between characters as there are when a teletype operator taps at her keyboard. Synchronous transmission is therefore of value when one communication line has several different terminals operating on it. In order to permit synchronous transmission, however, terminals must have buffers, and thus they are more expensive than asynchronous devices.

The synchronization of the transmitting and receiving machines on many systems is controlled by oscillators. Before a block is sent, the oscillator of the receiving machine must be brought exactly into phase with the oscillator of the transmitting machine. This step is done by sending a synchronization pattern or character at the start of the block. If this were not done, the receiving device would not be able to tell which bit received was the first bit in a character, which the second, and so on. Once the oscillators at each end are synchronized, they will remain so until the end of the block. Oscillators do, however, drift apart very slightly in frequency. This drift is very low if highly stable oscillators are used, but with those low enough in price to be used in quantity in input-output units it is significant. Most data-processing machines resynchronize their oscillators every few seconds for safety. Synchronization can also be maintained by "framing" blocks and carrying timing information in the frames.

On some systems, this process places an upper bound on the block length, though this is not always so because resynchronization characters may be sent in the middle of a block. The IBM range of "binary synchronous" equipment, for example, inserts two synchronization characters into the text at one-second intervals. In the U.S. ASCII code with parity checking (Figure 6.6), these would be coded 01101000 01101000. The receiving station is constantly looking for these characters and ensures that the transmitter and receiver are in step.

In addition to giving faster transmission because no START and STOP bits are needed between characters, synchronous transmission permits higher-speed modulation techniques to be used.

Synchronous transmission can give better protection from errors. At the end of each block, an error-checking pattern is transmitted. The coding of this pattern is selected to give the maximum protection from noise errors on the line. In addition to the error code at the end of the block, each character may also have a parity bit for checking. This, however, is often not done, and an end-of-block check is used alone. As discussed elsewhere, a parity bit is not too useful as a protection against communication line errors. It is an extremely useful protection against loss of bits in the core of a computer, because there it is likely that only one bit will be lost at a time. On a communication line, however, several bits are often lost at once because of a noise impulse or drop out. Where 2, 4, or 6 bits are changed in a character, a parity check will not detect this. Some form of block check that can detect the loss of several consecutive bits is therefore desirable. The faster the transmission, the more likely is the loss of more than one adjacent bit, and thus the block check becomes more important with faster transmission. Start-stop transmission may also have a block check.

BLOCK STRUCTURE

A block of bits sent by synchronous transmission must have certain features. It must, for example, start with the *synchronization pattern* or character. It will normally end with an *error-checking pattern* or character. The block length, as with other data records used by computers, may be of fixed length

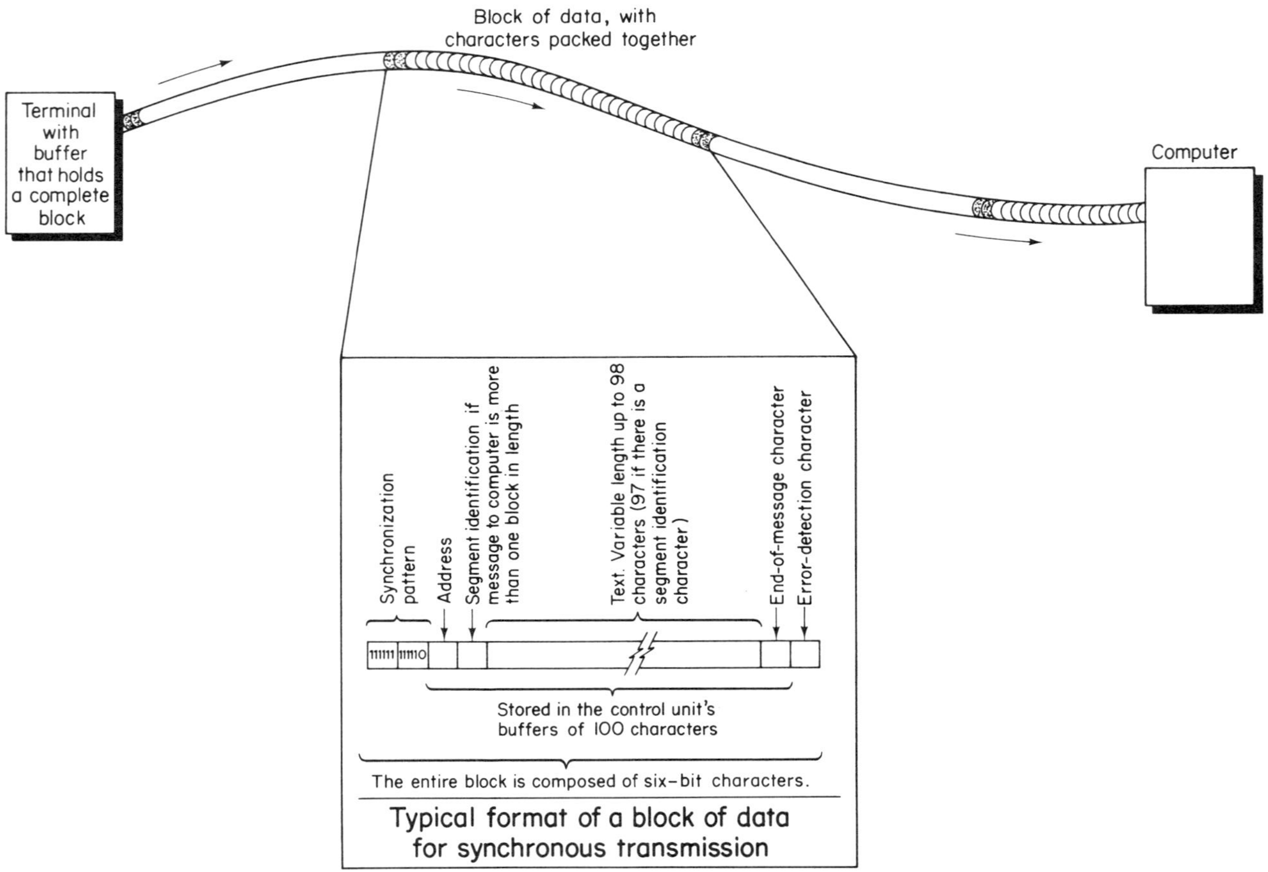

Figure 7.6. Synchronous transmission.

(Above) The UNISCOPE 300 Visual Communications Terminal uses synchronous transmission at 250 or 300 characters per second over a voice line.

(Left) The IBM 2770 Data Communications System. A variety of different devices can be attached to a control unit that is attached to a voice line and uses "binary synchronous" transmission.

The UNIVAC 1557/1558 Graphic Display Subsystem can either be connected directly to a computer or operated remotely.

Figure 7.7. **Machines That Use Synchronous Transmission.**

or variable length. It is often the latter, for this usually allows better line utilization. On most systems it would be necessary to pad many blocks with blank characters if fixed-length blocks were used. If the block is of variable length, an end-of-block pattern must be used to tell the receiving machine to begin the actions needed when a block ends. This pattern will normally be sent immediately prior to the error-checking pattern.

Often data are sent in the form of characters or groups (usually) of 6, 7, or 8 bits. The preceding patterns can be one, two, or more characters. The block in Figure 7.6, for example, uses six-bit characters. These are transmitted without parity checking, so the whole block is divided up into groups of six bits. The block must start with the following characters: 111111, 111110 in that sequence. This constitutes the synchronization pattern. A circuit in the receiving machine spends its life scanning the input for this pattern. When it finds it, then the receiving device knows that next bit it receives is the first data bit. The synchronization pattern is unique. The coding of characters must be such that it could not occur anywhere else in the transmission.

The block ends with a six-bit error-checking pattern (one character) and immediately preceding that is the end-of-message character. When the text is being transmitted, the receiving device is generating its own error-checking pattern, computed from the characters received. At the same time it is examining each character received to see whether it is the end-of-message character. When this is received, the machine knows that the next character is to be the transmitted error-checking pattern, and so it compares that with the pattern it has generated itself. If there is a difference, the receiving machine sends a message to the transmitting machine to demand a retransmission of that message.

After the synchronization pattern in each block comes the address of the terminal to which the message is going or from which the message has come. It is possible that messages transmitted to the computer may be longer than the maximum length of a block. In this case, they are divided into as many blocks as are necessary, and a character is used as a segment identifier to link them. The control unit places this, if it is needed, in the block immediately before the text. The text itself is again in six-bit characters and can be of any length up to 98 characters. This maximum is imposed by the size of the buffers, 100 characters, used in the control units in this example.

There are many variations of this type of format. Sometimes one character is designated as the "synchronization character," and a stream of these characters is sent continuously between messages when the line would otherwise be idle. At least two such characters are necessary prior to a message to establish synchronization.

Sometimes a block sent by synchronous transmission contains a more complex header. Sometimes it contains several records, all of variable length. Special characters must then be inserted to say where the header and text start and to indicate the end of the records and the end of the block. The end of the block is not necessarily the end of transmission. The sending device may have more data yet to come, and so an end-of-transmission character may also be needed in the repertoire. We thus have five demarkation characters:

1. Start-of-header
2. Start-of-text
3. End-of-record
4. End-of-block
5. End-of-transmission

SYNCHRONIZATION REQUIREMENTS

Basically, three levels of synchronization are needed in digital data transmission. These are needed regardless of the transmission mode. They are *bit synchronization, character synchronization*, and *message synchronization.*

Bit synchronization ensures that the receiving machine knows at what instant a bit starts and ends. The receiving machine must "sample" the bit at its center, not during a period of transition. *Character synchronization* ensures that the receiving machine knows which bit is which in a character. Without this synchronization the receiving machine might think the second bit of a character transmitted was in fact the beginning of the character, for example, so the characters would be misinterpreted. *Message synchronization* is needed to ensure that the receiving device knows which characters are the starting and ending characters of records or messages.

This is, in effect, saying that three types of timing information are needed by the receiving machine: timing information telling it the exact position of a bit, information as to which is the first bit of a character, and information as to what is the start and end of a message.

Bit synchronization may be all taken care of in the modem. Some modems are described as being "self-clocked" and automatically establish the "bit phase." The modem tells the data-processing machine when to sample the bit. Such is not always the case, however. Some modems merely send the data-processing machine a pulse train, and this machine must decide when to sample the pulses. In this case, a synchronization pattern is needed to establish the bit phase as well as the character synchronization.

The character synchronization pattern generally needs to be sent along with the bit synchronization pattern, if one is needed, at the start of each new message or data block.

Character synchronization is achieved in synchronous transmission by transmitting a unique bit pattern. In Fig. 7.6, for example, the pattern 111111–111110 is unique and when it is recognized, the receiving machine knows that the next bit will be the first bit of the first character (the address character). Sometimes a train of special identical synchronization characters are sent for this purpose. In IBM's "binary synchronous" transmission the synchronization character is 01010101. At least two of these are needed to achieve synchronization, and during idle periods the synchronization characters are sent continuously.

With start-stop information, each character in effect carries its own synchronization information. When the transition to the start bit is detected, this causes the receiving machine to sample the data bits at the correct intervals afterward. The starting and stopping give a form of transmission less resilient to noise and jitter than synchronous transmission. This is increasingly so at higher speeds.

Message synchronization is needed on start-stop or synchronous devices and is likely to take the same form with either of these. The *start-of-header*, *start of text*, *end-of-record*, *end-of-block*, and *end-of-transmission* characters discussed above, or their equivalents, may frame asynchronous as well as synchronous messages. Their use will be discussed further in chapters on line control.

HIGH-SPEED PULSE STREAM

A technique that might be regarded as a hybrid between synchronous and asynchronous transmission is one that uses a high-speed pulse stream traveling in a constant and unbroken fashion, to carry data in which the characters originate at random. Such a scheme is used in some in-plant systems. It has the advantage that both the terminals and the in-plant communication links can be relatively inexpensive.

As we commented in Chapter 3, bits can be carried at a very high rate over a simple pair of wires if regenerative digital repeaters (Figure 3.6) are used with sufficiently close spacing. The bit rate can, in fact, be so high that the bits can be used in a wanton fashion if this results in low-cost terminal equipment.

The wires may be arranged in a loop, as in Figure 7.8, that originates at a computer or controlling machine, wanders around various data-handling machines, such as those in Figure 7.9, and then returns to the computer. It may carry a train of small fixed-length blocks that convey characters to or from the terminal devices. Each block is able to carry one character. If a terminal at a particular instant has a character to transmit, its controller waits for an empty block, then places the character in it, along with the terminal address. Similarly, it receives characters being sent to the devices.

The reader might picture this as being like a never-ending train of railroad cars. Each car can carry one character. The station looks at the *station address* on each car (second byte in Figure 7.8) to see if that car is carrying a character for it. If so, it takes the contents of that car, which say what the character is to be used for and which terminal at that station it is intended for. Similarly, if the station has a character to send to the computer, it looks for the next *empty* car. It places the character into that car along with the station address, the terminal address, and a control character that says what the character is to be used for.

The terminals can transmit continuously or sporadically. The train will always be going by.

SUMMARY The various transmission modes are summarized in the appendix, on page 202, along with their relative advantages and disadvantages.

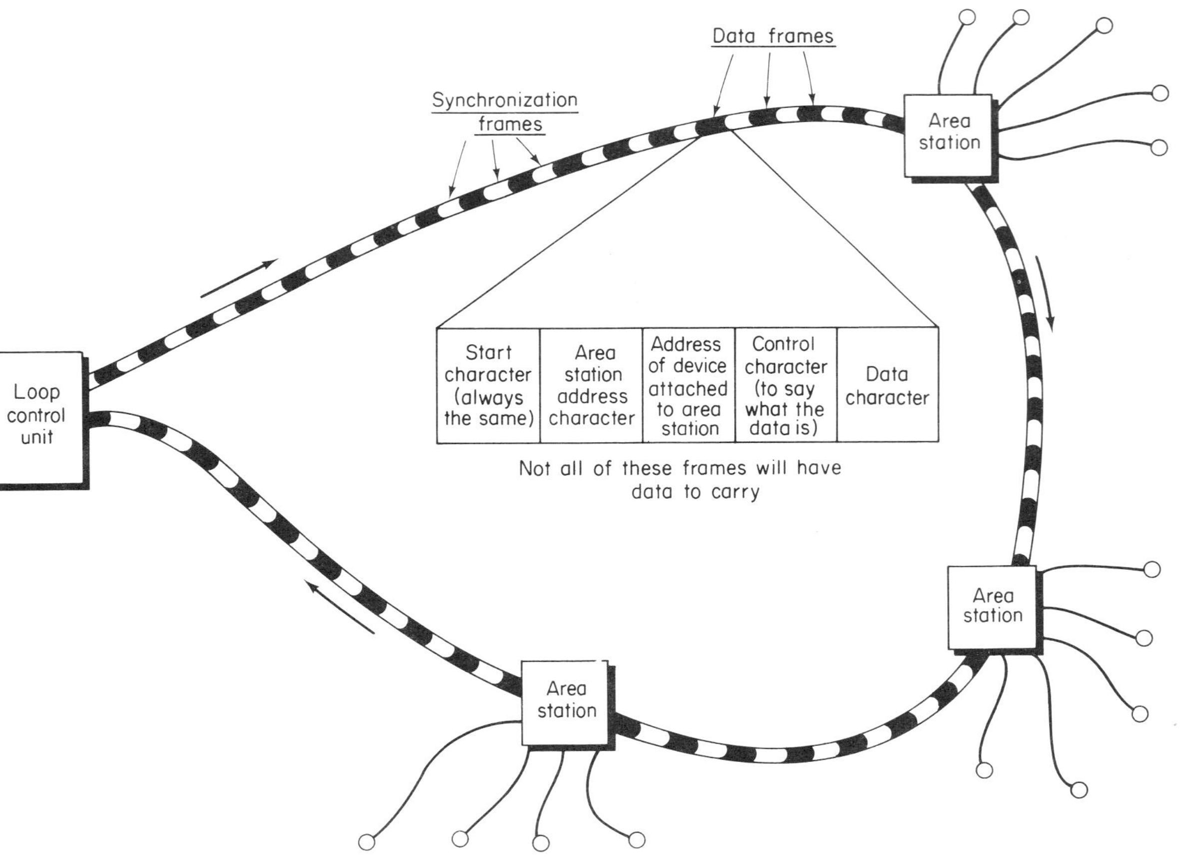

Figure 7.8. High-speed loop transmission.

The area station of the IBM 2790 system. The area stations are connected to a high-bit-rate in-plant loop as in Fig. 7.8.

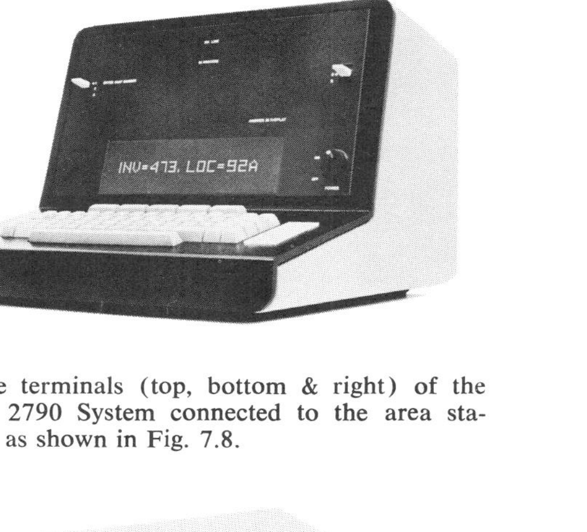

Three terminals (top, bottom & right) of the IBM 2790 System connected to the area stations as shown in Fig. 7.8.

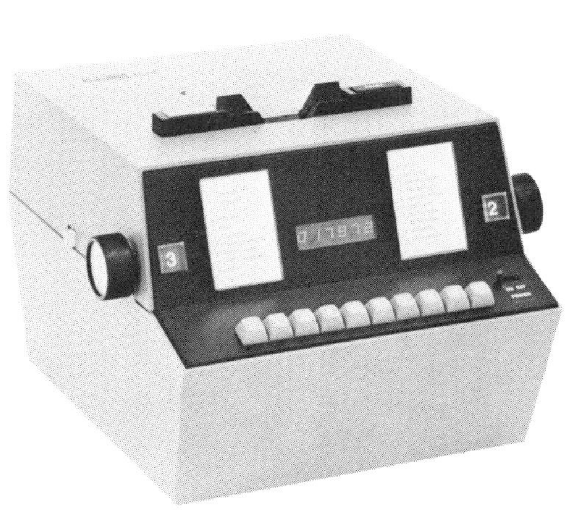

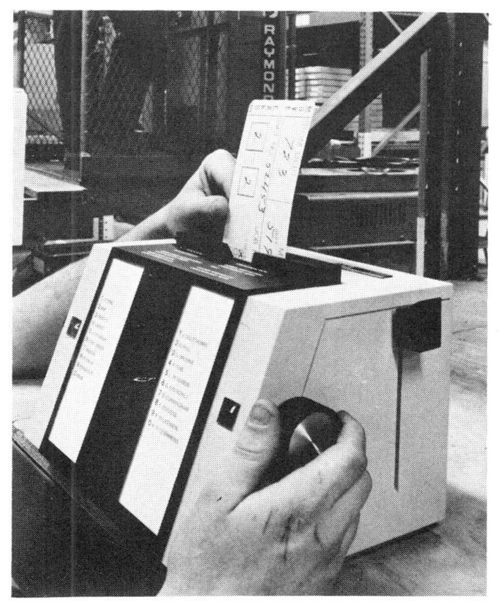

Figure **7.9. Machines attached to the high-speed loop of Fig. 7.8.**

8 LINE ERRORS AND THEIR TREATMENT

Errors occur when data are transmitted over telecommunication lines. Noise on the lines can destroy bits, switch a 1 bit to a 0, and vice versa. In addition to the continuing background of thermal noise, there are sharp noise *impulses*, occasionally of high magnitude. They are caused by ill-protected switching equipment, cross talk, pickup from electrical cables, atmospheric static, and a variety of other factors. If you listen down a telephone line when nobody is talking, you can occasionally hear crackles, hums, clicks, and whistles. I once heard violin music. These stray noises are usually of low intensity but occasionally are loud enough to damage the data being sent.

NUMBERS OF ERRORS

On a good-quality telephone line with conservatively designed modems, a typical error rate would be one bit incorrect in 100,000. If the modems are designed to maximize the data rate, the receiving machine has to recognize smaller changes in signal condition and so is more prone to misinterpret noise conditions and cause errors. Some typical error rates are cited in Table 8.1.

The figures for data rates above 2400 bits per second on a voice line are subject to fairly wide variation because the modulation techniques used become far more sensitive to noise and distortion than at the lower speeds.

When communication lines are constructed especially for data, much lower error rates than these are achieved. One error in 10 million bits or 100 million bits is more common. Unfortunately, most of our systems today must be linked together using lines that were not designed with the transmission of computer data in mind. The high error rate is part of the price we pay for the compromise of using lines intended for something else.

Table 8.1

Type of Channel	Transmission Rate (bits per second)	Bit Error Rate
Public voice line	600	1 in 500,000
	1200	1 in 200,000
	2400	1 in 100,000
	4800	1 in 10,000 to 100,000*
	9600	1 in 1,000 to 10,000*
Telex	50	1 in 50,000
Subvoice-grade lines	45 to 200	1 in 100,000

*before error-correcting codes are applied.

There are, however, steps that we can take to make the communication channel *appear* as error free as any other computer channel, discussed below.

THE EFFECT OF ERRORS

On many data transmission systems, the control of errors is of vital importance. On some, it is not of great significance.

Some systems transmit vital information, like accounting data, financial figures, military orders, encoded medical data or programs. These must be word perfect (or bit perfect). Other systems transmit the information that an operator is keying in at a terminal. The operator is likely to make far more errors than the data transmission line, especially if he is an unskilled one-finger operator. In this case, there is no point in worrying too much about the line—we must protect our system from the operator. Accuracy controls can be devised for human input [1]. On many systems a tight network of controls is necessary to stop abuse or embezzlement. It is also important to ensure that nothing is lost or double-entered when hardware failures occur on the system or when switchover takes place.

CUMULATIVE FILE ERRORS

In designing a computer system, it is important to know what error rate is expected. Calculations should then be done to estimate the effect of this error rate on the system as a whole. On some systems the effect of infrequently occurring errors is cumulative, and it is in situations such as these that special care is needed in eliminating errors. For example, if messages cause the updating of files, and an error in the message causes an error to be recorded on the file, then it is possible on some systems that as the months pass the file will accumulate a greater and greater number of inaccuracies.

Suppose, for example, that teletype transactions with one bit in 100,000 in error update a file of ten thousand records. Suppose that, on the average, an item is updated 100 times a month and that if any one of 20 five-bit characters is in error in the transmission, then the item will be updated incorrectly. After six months no less than 4500 file records will be incorrect. If an error-correction procedure on the telecommunication lines reduces the rate of undetected errors to one bit in 10 million, then 60 of the file records are likely to be incorrect at the end of six months. With one bit in 100 million, six records are likely to be wrong. Probability calculations of this type need to be done on various aspects of the system when it is being designed [2].

The systems analyst might use, as a rule of thumb, the error figures in Table 8.1 for doing probability calculations that will indicate to him the viability of different systems approaches. With them he may answer such questions as what degree of error checking is needed and what block length he should use for transmission.

POSSIBLE REMEDIES

A number of approaches can be used to deal with noise on transmission lines. All the approaches discussed below are found on data transmission systems in use today.

The first, and easiest, approach is to ignore the noise, which is often done. The majority of telegraph links in operation now, for example, have no error-checking facilities at all. Part of the reason is that they normally transmit English-language text that will be read by human beings. Errors in English language caused by the changing of a bit or of a small group of bits are usually obvious to the human eye, and we correct them in the mind as we read the material. Telegrams that have figures as well as text in them commonly repeat the figures. This inexpensive approach is also taken on computer systems where the transmission handles verbal text. For example, on administrative message-switching systems, it is usually acceptable to have transmission to and from unchecked telegraph machines. If the text turns out to be unintelligible, the user can always ask for a retransmission.

An error rate of one bit in 100,000 is possibly not quite so bad as it sounds. Suppose that we consider transmitting the text of this book, for example, and coded it in five-bit Baudot code. If one bit in 100,000 were in error, in the entire book there would be about 40 letters that were wrong. The book would certainly still be readable, and the majority of its readers would not notice most of these errors. The human eye has a habit of passing unperturbed over minor errors in text. This book was first set in galley proofs by the compositor. The proofs were checked by a proofreader. It was then divided into pages and page proofs were produced. By now most of the errors should have been removed from the text, but, in fact, those remaining correspond to an error rate of one bit in 20,000, an error level higher than that which would be found on unchecked telegraph transmission.

In any case, on most systems *some* of the errors are ignored. An error-detection procedure that catches all of them has been too expensive. Many systems in current use might raise the level of undetected errors from one bit in 100,000 to one bit in 10 or 100 million. It is possible to devise a coding scheme that gives very much better protection than this. In fact, one that is on the market gives an undetected error rate of one bit in 10^{14} (that is, 1 in 100,000,000,000,000), but this is expensive.

An error-detection rate of one bit in 10^{14} is much better than is needed for most practical purposes. If one had transmitted nonstop at that rate over a voice line at 2400 bits per second, for a normal working week, (no vacations) since the time of Christ, one would probably not have had an error yet!

By using sufficiently powerful error-detecting codes, virtually any measure of protection from transmission errors can be achieved. The codes necessary to achieve this are described in Chapter 5 of the author's *Teleprocessing Network Organization.* It is hoped that some future terminals will give a very high measure of protection at reasonable cost, using "large-scale integration" circuits for encoding and error detection. Because of the rapidly decreasing cost of mass-produced logic circuits, the cost of virtually perfect error detection is falling to a reasonable level.

DETECTION OF ERRORS

To *detect* communication errors, redundancy is built into the messages transmitted. In other words, more bits are sent than need be sent for the coding of the data alone.

Redundancy can be built into individual characters. This is done by using parity checks. It is also done by using certain character codes such as a 4-out-of-8 code (Figure 6.5). As we commented before, the parity check is not very effective in data transmission (although often used) because a burst of noise frequency destroys more than one bit in a character. The higher the transmission speed, the more likely this is to be so. A noise impulse 1.5 milliseconds in duration, for example, may destroy only one bit if the transmission rate is 600 bits per second, but at 1200 bits per second it may wipe out 2 bits; at 4800 bits per second 4 or 5 bits; and so on. If an even number of bits are destroyed, the error will not be detected.

The 4-out-of-8 code and other such character codes are a little safer because, to be undetectable, errors must be compensating; in other words, if a 1 bit is changed into a 0, then a 0 bit must also be changed into a 1. This sometimes happens with the bursts of noise and oscillating effects on communication lines. It is more likely to happen at higher bit rates, and with high-speed modulation techniques.

Rather than using redundancy within a *character* in this way, it is better to use redundancy within a *block* of characters. A block of many characters will be followed by a cluster of error-checking bits, perhaps in the form of one or more error-checking characters. A block of this structure was shown in Figure 7.6, and may have more than the one error-detection character shown. Some transmission schemes have character checking—for example, a parity bit on each character—in addition to the check on the block. It can be shown, however, that much more efficient error detection results from using all the redundant bits for block checking, not character checking.

The probability of transmission errors remaining undetected can be made very low, either by a brute force method that gives a high degree of redundancy or, better, by an intricate method of coding. Many systems in actual use do not employ a *very* secure error-detection scheme but compromise with one that is not too expensive but that will probably let one error in a thousand or so slip through. In the future, it is probable that an increasing number of transmission devices will use an error-detection scheme of high quality.

DEALING WITH ERRORS

Once the errors have been detected, the question arises: What should the system do about them? It is generally desirable that it should take some automatic action to correct the fault. Some data transmission systems, however, do not do so and leave the fault to be corrected by human means at a later time. For example, one system that transmits data to be punched into cards causes a card to be offset in its position in the stacker when an error is detected. The offset cards are later picked out by the operator, who then arranges for retransmission. In some real-time terminal systems automatic retransmission has not been used because it is easy for the terminal operator to reenter a message or request retransmission. In general, it is much better to have some means of automatic retransmission rather than a manual procedure, and it is usually less expensive than employing an operator for this purpose.

Again, on some systems, it is possible to ignore incorrect data, but it is important to know that it *is* incorrect. On such systems error detection takes place with no attempt to correct the errors. This could be the case with statistical data where erroneous samples can be discarded without distortion. It is used on systems for reading remote instruments where the readings are changing slowly and an occasional missed reading does not matter. The advantage of a detection-only scheme is that it requires a channel in only one direction. In systems with telephone and telegraph lines, this is not a worthwhile advantage because such channels are half or full duplex. However, it is a great disadvantage with certain tracking and telemetry systems, and here we find detection-only schemes. Clearly, for most commercial applications, they will not suffice.

ERROR-CORRECTING CODES

Automatic correction can take a number of forms. First, sufficient redundancy can be built into the transmission code so that the code itself permits *automatic error correction* as well as detection. As no return path is needed, this is sometimes referred to as *forward-error correction.* To do this effectively in the presence of *bursts* of noise can require a large proportion of redundancy bits. Codes that give safe forward-error correction are, therefore, inefficient in their use of communication line capacity. If the communication line permitted the transfer of information in one direction only, then they would be extremely valuable. However, again, most earthbound systems use half-duplex or full-duplex communication links. In general, error-*correcting* codes alone on voice-grade or subvoice-grade lines do not give us as good value for money, or value for bandwidth, as error-*detecting* codes coupled with the ability to retransmit automatically data that are found to contain an error. Forward-error correction becomes advantageous when the number of errors is so high that the retransmission of data necessary would substantially degrade the throughput. On higher-speed links, the argument for forward-error correction becomes stronger because the time taken for reversing the direction of transmission is equivalent to many bits of transmission. This time is relatively high on wideband links and on half-duplex, voice-grade links with high-speed modems.

Where forward-error correction is used—for example, on high-speed modems—it is often backed up with more secure error-detection techniques, but with the advantage that less data have to be retransmitted.

LOOP CHECK

One method of detecting errors does not use a code at all. Instead, all the bits received are retransmitted back to their sender, and the sending machine checks that they are still intact. If not, then the item in error is retransmitted. Sometimes referred to as a loop check or echo check, this scheme is normally used on a full-duplex line or on a continuous loop line. Again, it uses the channel capacity less efficiently than would be possible with an error-detection code, although often the return path of a full-duplex line is underutilized in a system, for the system does not produce enough data to keep the channel loaded with data in both directions. A loop check is most commonly found on short lines and in-plant lines where the wastage of channel capacity is less costly. It gives a degree of protection that is more certain than most other methods.

RETRANSMISSION OF DATA IN ERROR

Many different forms of error *detection* and retransmission are built into data-handling equipment. In a typical high-speed paper-tape transmission system, a "vertical" parity check—that is, a parity check on each character—is used along with a "horizontal" checking character at the end of a block of characters. At the end of each block, the receiving station sends a signal to the transmitting station saying whether the block has been received correctly or whether an error has been detected. If any error is found, both the transmitting tape reader and the receiving tape punch go into reverse and run backward to the beginning of that block. The punch then erases the incorrect data by punching a hole into every position—the *delete character* code.

The block is then retransmitted. If transmission of the same block is attempted several times (four times on much equipment) and is still incorrect, then the equipment will stop and notify its operator by means of a warning light and bell or buzzer. Other automatic facilities are usually used to detect broken or jammed paper tape and to warn when the punch is running short of tape.

Where data are being transmitted to a computer, automatic retransmismission is sometimes handled under control of the program, and sometimes by circuitry external to the main computer. More detailed examples of this will be given later.

HOW MUCH IS RETRANSMITTED?

Systems differ in how much they require to be retransmitted when an error is detected. Some retransmit only one character when a character error is found. Others retransmit many characters or even many messages.

There are two possible advantages in retransmitting a *small* quantity of data. First, it saves time. It is quicker to retransmit 5 characters than 500 when an error is found. However, if the error rate is one character error in 50,000 (a typical figure for Telex and telegraph lines) the percentage loss in speed does not differ greatly between these two cases. It *would* be significant if a block of 5000 had to be retransmitted.

Second, when a large block is retransmitted, it has to be stored somewhere until the receiving machine has confirmed that the transmission was correct. This is often no problem. In transmitting from paper tape, for example, the tape reader merely reverses to the beginning of that block. The paper tape is its own message storage. The same is true with transmission from magnetic tape or disk. With transmission from a keyboard, however, an auxiliary storage, or *buffer*, is needed if there is a chance that the message may have to be retransmitted automatically. On some input devices, a small core storage unit constitutes the buffer. On others, the keys themselves are the storage. They remain locked down until successful transmission is acknowledged. Again, several input devices may share a common control unit, and this contains the buffer storage. Buffer storage in quantity can be fairly expensive, so it may be better to check and retransmit only a small number of characters at a time. Again, when data on punched cards are transmitted, a buffer is needed for retransmission unless the machine is designed in an ingenious manner so that the card can be reread if required.

The *disadvantages* of using small blocks for retransmission are first that the error-detection codes can be more efficient on a large block of data. In other words, the ratio of the number of bits *with* the error-detection code to the number if the data were sent *without* protection is smaller for a given degree of protection if the quantity of data is large. Second, where blocks of data are sent synchronously, a period of time is taken up between blocks in control characters and line turnaround procedures. The longer the block, the less significant this wasted time.

The well-designed transmission system achieves the best compromise between these factors.

Let us consider some examples of different retransmission quantities:

1. *Retransmission of one character.* Characters are individually error checked, as with a 4-out-of-8 code. As soon as a character error is detected, retransmission of that character is requested. This is likely to be used only on a very slow link.
2. *Retransmission of one word.* The British ICL Type 7000 series equipment uses blocks of nine characters. Figure 8.1 shows such a block. The block has a parity bit for each "row" or character, and also for each column. A transmitting terminal has buffers that hold two such blocks. Should the receiver detect an error in row or column parity, retransmission of that block is requested.
3. *Retransmission of a message or record.* The above "word" was both short and fixed in length. Many machines retransmit complete messages or records at one time that are much longer than the above word and that more often than not are variable in length. They may

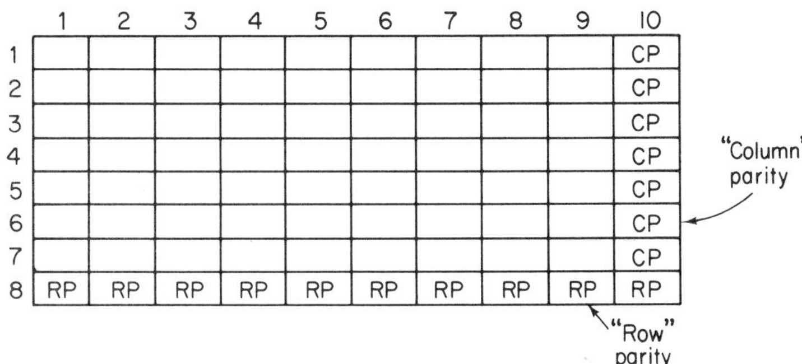

Figure 8.1

have a format such as that in Figure 7.7, in which an end-of-transmission character terminates the text and this is followed by the error-checking characters. The record may be retransmitted from backspaced tape from a variable-length area in the core of a computer or from a buffer in a control unit attached to the transmitting device. If a buffer is used, there may be a maximum size for the amount that can be retransmitted. If the message exceeds that size, it is broken into separate messages that are linked together with a control character indicating that a given transmission has not completed the message.

4. *Retransmission of a block of several messages or records.* When the transmission speed is high, it becomes economical on many systems to transmit the data synchronously in large blocks. These may contain not one "message" or "record," but several. On most systems, when an error occurs in any of the records, the whole block of records is retransmitted. By using machines with good logic capabilities it would be possible to resend only the faulty record and not all the other records in the block.
5. *Retransmission of a batch of separate records.* Sometimes a control is placed on a whole batch of records. Like the controls conventionally used in batch data processing, the computer adds up account numbers and/or certain data fields from each record to produce *hash totals*. These totals are accumulated at the sending and the receiving end and are then compared. At one of these stations such a control might, in some cases, be produced manually on an adding machine, and it is often used to detect not only the errors in transmission but also errors in manual preparation of data. When one computer sends a program to another computer, it is vital that there should be no

undetected error in the program, so the words or groups of characters are added up into an otherwise meaningless hash total. This hash total is transmitted with the program, and only if the receiving computer obtains the same total in *its* addition is the program accepted.

Some form of batch control of this type is often used, where applicable, *as well* as other automatic transmission controls overriding safety precaution. Its use is entirely in the hands of the systems analyst and can be made as comprehensive and secure as he feels necessary.

A typical application of batch totals might be on the network shown in Figure 8.2. Here transactions are to be sent from many branch locations to the head office computer. The transactions are punched into paper tape at the branches on telegraph machines. The transactions are grouped into batches of about 50 and these are added up on an adding machine before punching. The totals obtained are punched as the last transaction of the batch.

The batches are sent on telegraph lines to a programmed hold-and-forward concentrator. The machine handles all the data from one area and forwards it periodically to the distant head office computer on a dial-up voice line using synchronous transmission. The main purpose is to reduce the cost of the communication lines.

As the concentrator receives a batch, it adds up the batch totals and compares them with the totals received. It prints a message at the branch, saying whether the totals were correct or not. If this were not done, the telegraph transmission would be unchecked. Here the batch totals are being used to check the accuracy of punching as well as the accuracy of transmission. The concentrator stores the data on its file and later dials up the head office computer to transmit the batches it has collected. This link has other error-detection facilities, but for additional safety, the batch totals are again checked by the head office computer.

6. *Retransmission at a later time.* The batch totals in Figure 8.2 were checked immediately; the transmission was complete, and the sender was notified whether they were correct or not. Some forms of validity check might not be capable of being used until the items are processed. They may, for example, necessitate comparing transactions with a master tape. They are, nevertheless, valuable error controls, and an originating computer might keep the data in its files until a receiving computer has confirmed this validation.

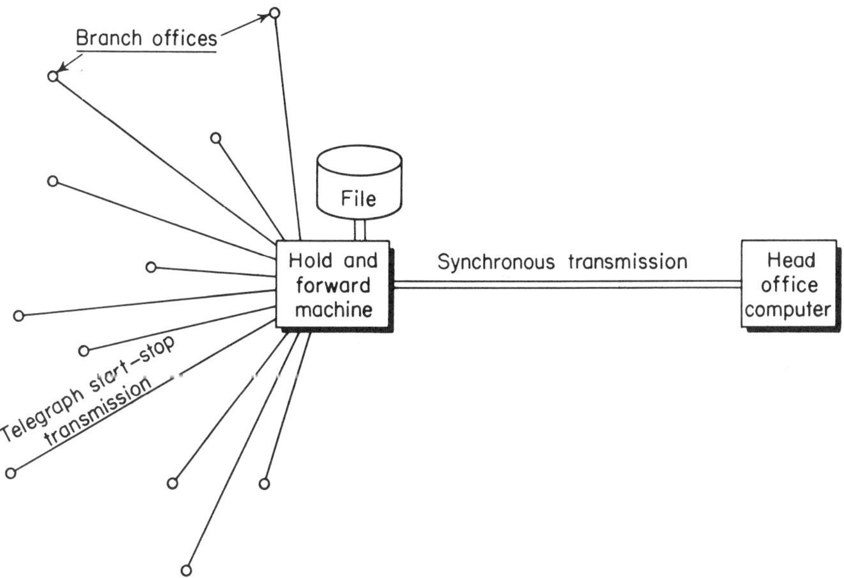

Figure 8.2. Batch error control is here performed at the hold-and-forward machine.

TRANSMISSION ERROR-CONTROL CHARACTERS

In order to govern the automatic retransmission of information in which an error has been detected, a number of special characters (*control characters*) are used, sometimes sequences of characters. The ASCII code, for example, uses the codes ACK, NAK, CAN, and DEL.

The ACK code is used by the receiver to signal the transmitter that a block of code has been received correctly. Similarly, the NAK code is sent by the receiving terminal to tell the transmitter that a block of code received had an error in it. When the transmitter sends a block of code on most systems, it waits before it sends the next one until the ACK or NAK control character is received from the transmitter. If ACK is received, it proceeds normally; if NAK, it resends the block in error.

The transmitter itself commonly does some error checking on what it sends. It is possible that the circuits doing this may detect an error in a message on which transmission has already begun. The transmitter must then cancel the message, and so it sends a CAN character. The DEL character is normally a character that should be totally ignored. It is used to delete characters already punched in paper tape, by punching a hole in every position.

The transmitting and receiving machines have circuits designed to detect these special characters.

ODD-EVEN RECORD COUNT

It is possible that the control characters themselves or end-of-transmission characters could be invalidated by a noise error. If this happens, then there is a danger that a complete message might be lost or two messages inadvertently joined together. It is possible that during the automatic retransmission process a message could be erroneously sent twice. To prevent these errors, an odd-even count may be kept of the records transmitted.

Sometimes, at the start of a block, a control character is sent to indicate whether this is an odd-numbered or even-numbered block. On some systems two alternative *start-of-transmission* characters are used. With other schemes, it is the ACK characters that contain this odd-even check. Two different ACK-type signals may be sent: ACK 0 and ACK 1. On the ASCII code there is only one ACK character, so if this is used a two-character sequence may be employed.

If an odd-numbered block does not follow an even-numbered block, then the block following the last correct block is retransmitted. It is very improbable indeed that two blocks could be lost together or that two blocks are transmitted twice in such a manner that the odd-even count would not detect the error. However, some systems use a *serial number* to check that this has not happened, instead of the odd-even count. The serial number may be examined by hardware, but often it is a check that is applied by a computer program; one of its main values is to bridge the continuity gap when a terminal, computer, or other hardware, failure occurs. During the period of recovery from such a failure, there is a danger of losing or double-processing a transaction.

MINIMUM NUMBERS OF BITS FOR ERROR RECOVERY

Error control need not have elaborate transmission of control characters back and forth. Some transmission schemes seem unnecessarily burdened by their use of control messages. Error control *could* be achieved with two control bits only, one for saying whether a message had been received correctly or not, and one as an alternation bit giving an odd-even count.

On multipoint lines it is difficult to recover from *addressing* errors with certainty by means of answerback schemes with control characters. The simplest method is a scheme with positive acknowledgments *only*. The receiver simply ignores incorrect messages. The transmitter sends a message with an odd-even block count and waits for acknowledgment of its correct receipt. If no acknowledgment is received after a specified time, then it resends the message, with the same odd-even count. One of two things might have gone wrong. First, the message might have been received with an error. In this case, it is resent as would be required. Second, the positive acknowledgment might have been destroyed. In this case, when the receiver receives a second copy, he will send the positive acknowledgment again. Providing that all errors are detected by error-detecting codes, this simple scheme will be infallible on point-to-point or multipoint lines.

9 MAN–COMPUTER DIALOGUE

In order to bring the power of computers and the information in their data banks to the maximum number of people, careful attention must be paid to the man–machine interface. Increasingly, in the next decade, man must become the prime focus of systems design. The computer is there to serve him, to obtain information for him and to help him do his job. The ease with which he communicates with it will determine the extent to which he uses it. Whether or not he uses it powerfully will depend on the man–machine language available to him and how well he is able to understand it. For the ordinary manager and for many other types of computer users, remarkably little has been done to provide an efficient man–machine interface as yet.

The interface is going to differ greatly from one man to another and from one machine to another. Different applications will need fundamentally different types of dialogue structures [1]. Some are very complex and need a high level of intelligence: others are simple. In some applications today, one can observe the "man-in-the-street" who has never touched a terminal before, sitting down at one and carrying on a successful, if simple, dialogue with a computer. On the other hand, one also finds terminals being thrown out a few months after installation because the intended user never learned to communicate successfully with the system.

The man–computer dialogue should be a starting point in the design of the data communication facilities in many real-time systems [2].

This chapter shows two segments of man–machine dialogue. They have been chosen as being typical in that they illustrate types of dialogue in common use. Both require a trained operator, not an inexperienced one, and in this sense neither perhaps, is a good example of the type of dialogue we should be constructing in future for the vast mass of largely untrained people we want to see using computers [1].

The first specimen is in Figure 9.1. It takes place at a teletype machine that responds at the frustratingly slow speed of 10 characters per second. The user is doing a simple calculation in order to plot a set of curves of the function

$$P(T) = e^{-[(1-\rho)\,T/S]}$$

In both examples the operator's input is shown in **heavy type** and the computer response is shown in *italics.*

In a dialogue such as that in Figure 9.1, the operator is using a programming language—in this case BASIC. The first action is to inform the computer that BASIC is the language to be used. The operator makes several mistakes as he keys in statements. Some he notices and corrects himself. Some the computer catches, and tells him. He obtains his results much quicker than with a desk calculator or with a conventional non-time-sharing computer.

The second example relates to an airline agent not trained in programming. She converses with the computer, using predetermined mnemonics, and for such cases, it is better to use the term "dialogue" rather than "language" because language suggests a formal computer language (like BASIC).

In this case, the agent has a visual display unit and so is not frustrated by slow responses as she would be if she used a 10-characters-per-second teletypewriter as in the previous case. She often has a potential customer on the telephone and needs to be able to act promptly, obtaining quick responses from the computer.

To assist in the dialogue, some keys on her terminal have been especially labeled.

The dialogue proceeds in 12 steps as follows:

1. A Mr. Goldsmith telephones the airline to say that he wants to modify his reservation. The agent he speaks to instructs her terminal to display his record—Flight 21 on the 25th of March. She types (Figure 9.2):

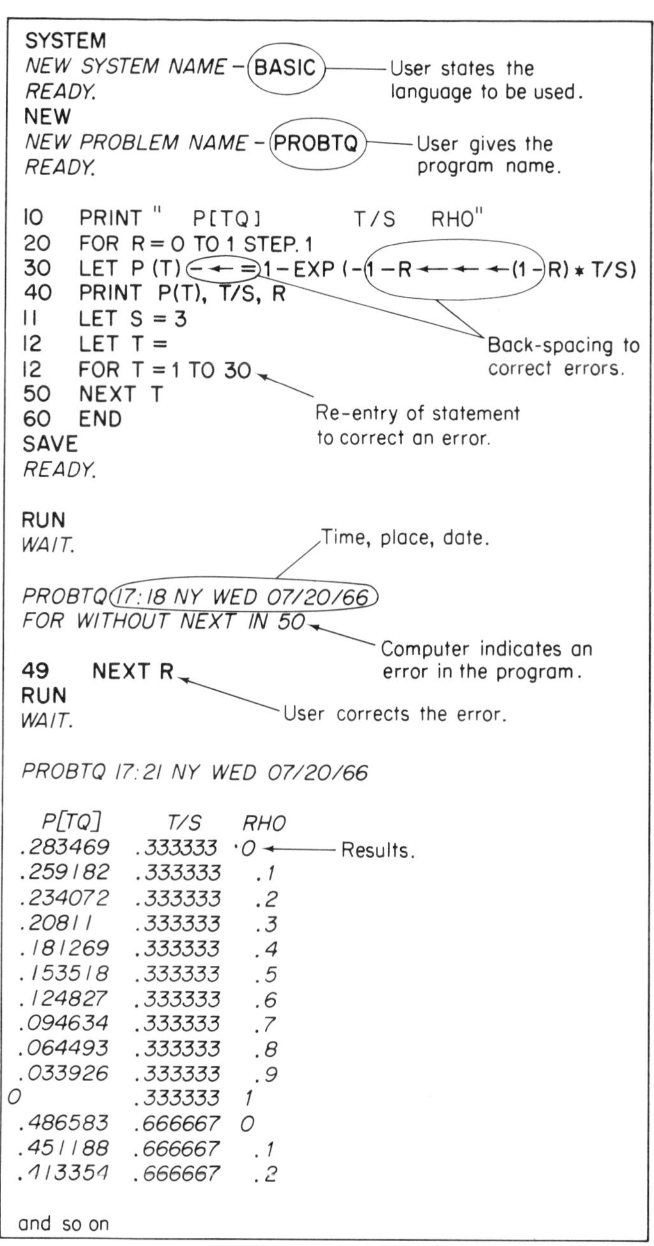

Figure 9.1. A calculation performed at a teletypewriter with the BASIC language.

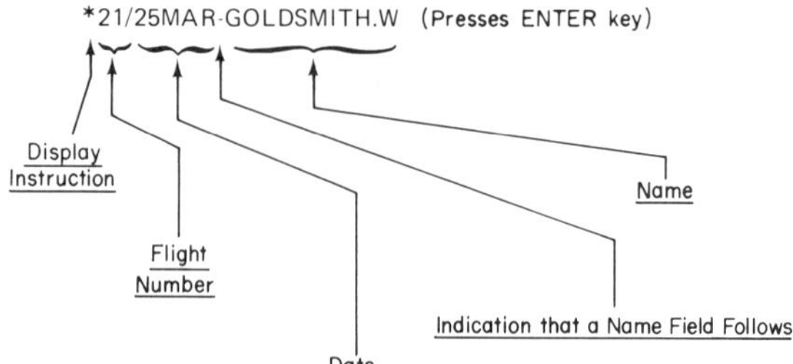

Figure 9.2

2. The computer displays details of his record on the screen of the terminal (Figure 9.3):

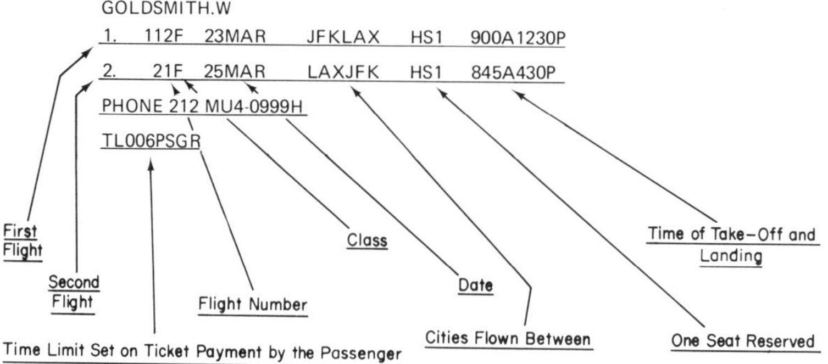

Figure 9.3

3. The operator checks that this is indeed the required record, and the passenger indicates that he wishes to change the second flight of his journey. He wants to fly back from Los Angeles (LAX) to New York's Kennedy Airport (JFK) on the 26th of March rather than the 25th.

   The girl informs the computer that the second flight of the journey is to be changed. She types (Figure 9.4):

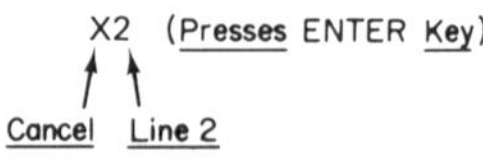

Figure 9.4

(The X key is labeled "Cancel.")

4. The computer responds:

NEXT SEG ENTRY REPLACES 2

5. She types (Figure 9.5):

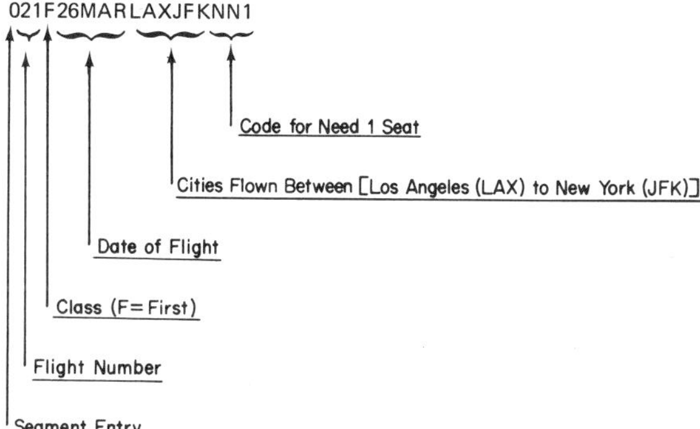

Figure 9.5

This statement is a request for a seat on the 26th of March rather than the 25th. "NN1" is a code meaning that the agent needs one seat.

6. The computer replies with a replacement for line 2 above:

2. 21F 26MAR LAXJFK HS1 845A430P

7. The operator attempts to book this seat by pressing the "E" key, which is labeled "END TRANSACTION."

E

8. The computer asks who requested this modification:

WHO MADE CHANGE?

9. The operator replies that the passenger himself requested it (Figure 9.6). The "6" key means "change made by . . .":

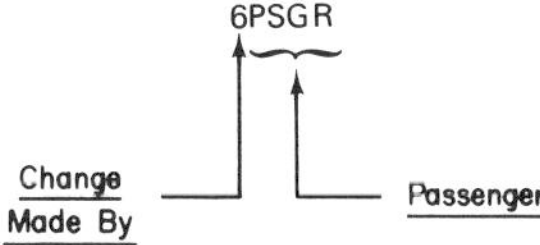

Figure 9.6

10. The computer indicates that this message has been received by placing an asterisk at the end of it.

6PSGR*

11. The operator again presses the "E" key, and this time the computer carries out the "END TRANSACTION" operation and updates the appropriate files.
12. It responds (Figure 9.7):

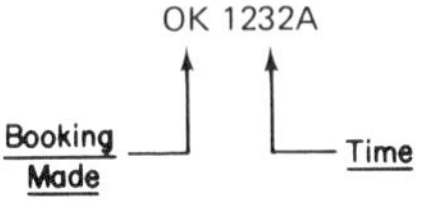

Figure 9.7

SPORADIC USE OF COMMUNICATION LINE

In the examples illustrated, the line tends to be used in a sporadic fashion. The user will often pause to read a response, or to think. When he types, he may do so slowly. This is usually the case.

Figure 9.8 gives a time scale showing the timing of the airline agent's dialogue. The terminal has a buffer. Because of this, the line is not occupied for the time the agent is keying data in but only for the time it takes to transmit that data. It will be seen from Figure 9.8 that the line is occupied for only a small proportion of the total time. Furthermore, there is no transmission from this terminal for the next two minutes.

In fact, a leased voice line to a terminal such as this is likely to transmit at 4800 bits per second. The occupancy of the line shown in Figure 9.8 assumes a speed of only 480 bits per second. This was done in order that the transmission bursts could be drawn at all. The reader should imagine that, in fact, they are one tenth of the thickness of those in the figure.

The line is being so drastically underutilized that it is clearly desirable to find some way to interleave the transmissions of many such terminals. In other words, there is a need for *multiplexing*.

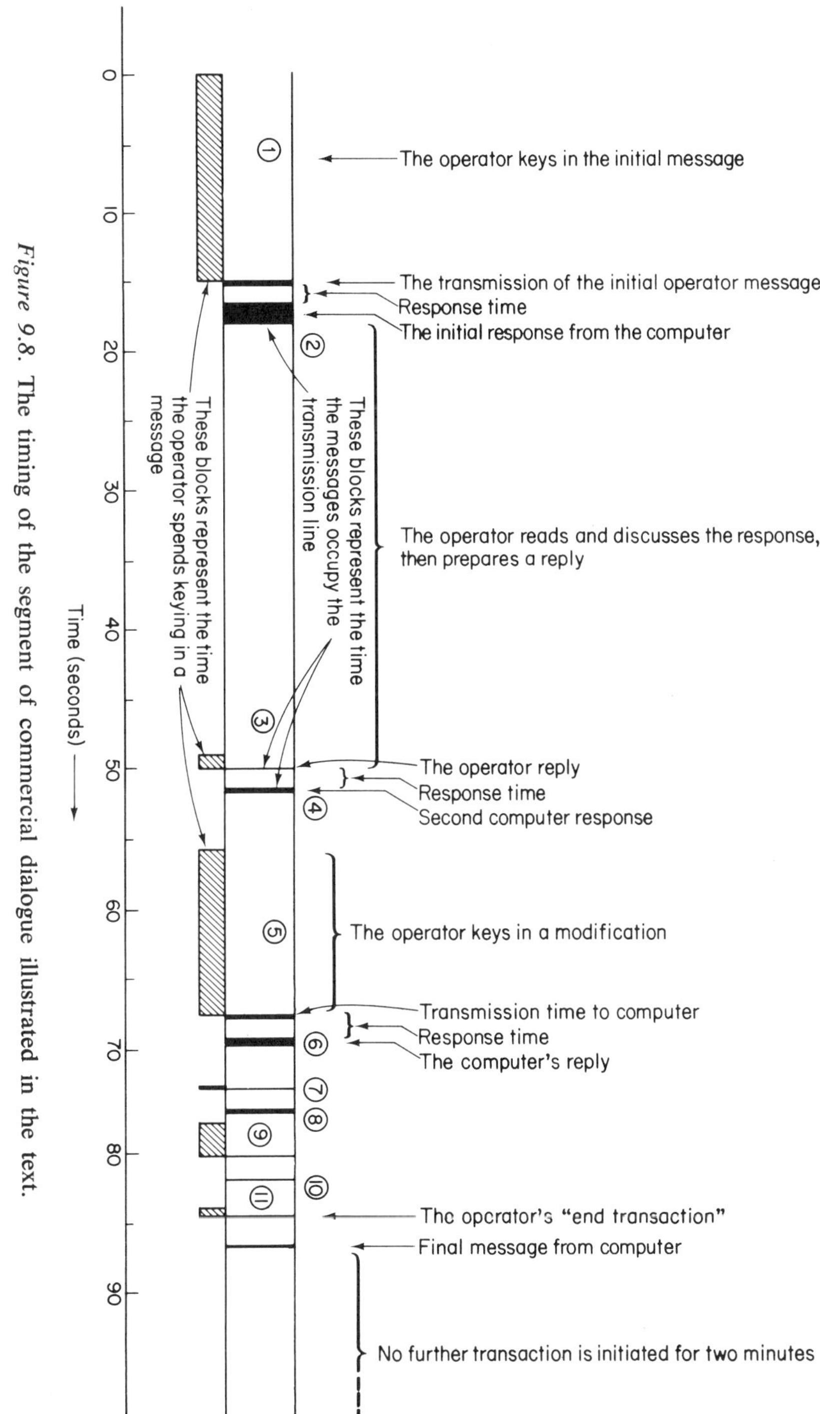

Figure 9.8. The timing of the segment of commercial dialogue illustrated in the text.

10 THE NEED FOR MULTIPLEXING

When writing this book I was at a conference in Europe in which we decided to demonstrate some points by remotely using a computer in the United States. A teletype machine was employed as the terminal, and the communication line, a telephone line, was obtained with no difficulty. At the end of half an hour's use I examined the print-out and estimated that about 3000 characters had been transmitted in total to and from the computer. This number is probably typical when using a teletype machine for time sharing.

The voice line was capable of transmitting 4800 bits per second with ease (a higher rate could be attained with sophisticated modems). In half an hour, then, it could transmit $1800 \times 4800 = 8{,}640{,}000$ bits, and in fact it had sent only $3000 \times 7 = 21{,}000$ bits of data. One could say that the efficiency of the way we had used it was $21{,}000/8{,}640{,}000 = 0.0024$—a very poor way to use such an expensive facility.

EFFICIENCY: 0.0001

Voice lines are now being constructed using digital (PCM) techniques in which one telephone channel becomes equivalent to 56,000 bits per second in each direction (see details of the Bell System T1 carrier in the author's *Telecommunications and the Computer*). This could potentially transmit 1800×56000 bits in half an hour, in each direction. The efficiency with the above use of time sharing could then be said to be

$$\frac{21{,}000}{2 \times 1800 \times 56{,}000} = 0.0001$$

In an industry with the logic capability of the computer industry, an efficiency of 0.0001 ought not to survive long. If we can push the efficiency up to 0.25, we have an improvement of 2500 times, and on the analog voice line used at 4800 bits per second, a one-hundredfold improvement. This can be done if we can arrange for different users to share the transmission capacity simultaneously. The computer, after all, was being time-shared. We must time-share the lines as well.

The need for sharing is even more pronounced with other transmission facilities. The microwave links, coaxial cables, and even wire pairs that form a nation's telecommunication links [1] can be made to carry *very* much higher bit rates than the preceding ones. The wire pair of the Bell System T1 carrier transmits a potentially usable 1,344,000 bits per second in each direction [2].

SHARING THE LINES

The sharing of transmission lines can be carried out in a variety of different ways, as we shall see. Basically, two problems are involved. The first is the technical problem of combining different transmissions on the same line. This is relatively easily solved. If several separate transmissions are sent over the same line *at the same time*, this is referred to as *multiplexing*. Many different *multiplexers* are available for data transmission and all achieve a high level of efficiency—0.8 to 0.9, for example.

The second problem is that of bringing together a sufficient number of users to fill the group of channels that has been derived from the line. Sometimes there are enough users within one organization. They can be brought together to share a leased line. Often this is not the case, however.

WAYS TO ORGANIZE LINE-SHARING

There are a number of ways in which the sharing of communication lines might be arranged.

1. *Different Users of One System*

On a system with many terminals, the communication network will be organized so that lines are shared between the terminals. A simple arrangement could be that shown in Figure 10.1. Here a group of terminal users do not have a computer in their building. They use a distant computer by means of a voice line that is shared between them.

Sometimes the terminal users are in different buildings, scattered geographically. In this case, unshared lines may connect them to a point at

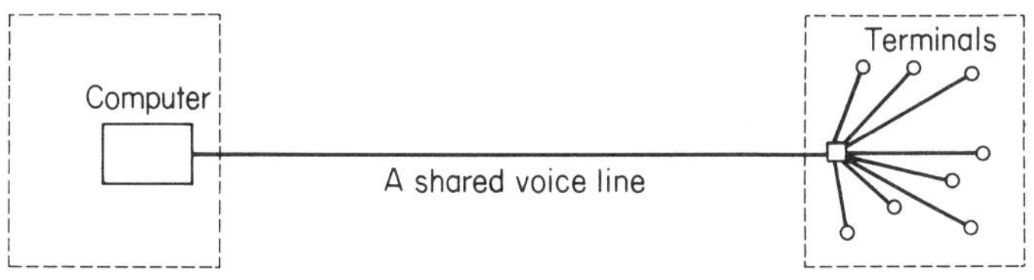

Figure 10.1

which sharing begins. A network of this type may link many users, as shown in Figure 10.2.

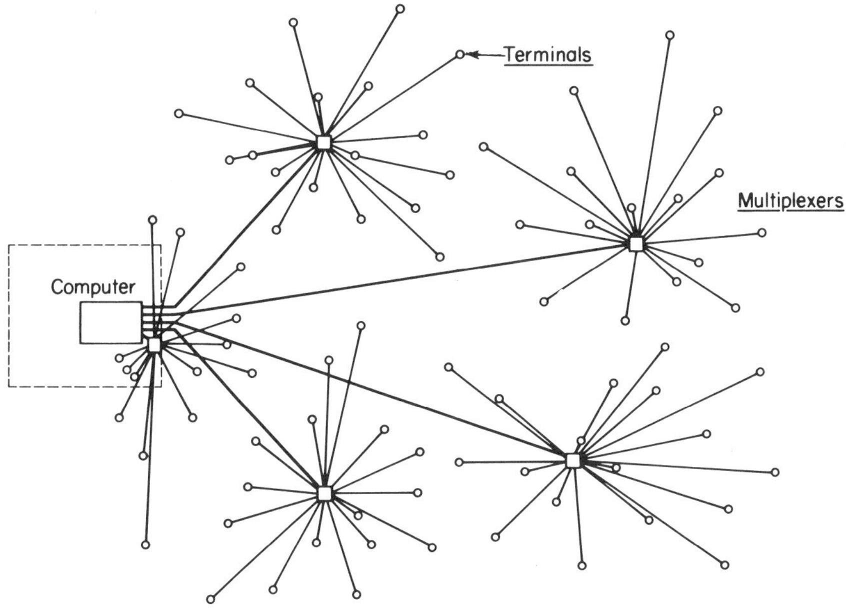

Figure 10.2

A variety of different devices and techniques to facilitate sharing are available. These are reviewed in the next chapter. It is the job of the systems analyst to select the best for his particular system, and this can be a complex job.

2. *Different System Users within One Organization*

Corporations that are up to date in their use of data processing have a proliferation of different systems. A data communication network that is shared by different system users may be set up within the corporation. In some cases this may be simply a facility that enables tie lines to be used for data. A leased TELPAK link may be split up by multiplexing equipment into many separate data links, used for separate purposes. The data links so derived need not necessarily be of the same data rate. A variety of such multiplexing equipment is on the market.

It is probable that complex private data networks will grow up in large corporations in the years ahead, using private switching facilities and multiplexers. Three types of switching are in use in this way, all of which will be described briefly in the next chapter. They are

(a) *Line switching*, in which circuits are switched as in the telephone network.

(b) *Message switching*, in which data messages are stored at switching centers and then relayed onward.

and

(c) *Packet switching*, in which data packets of fixed format travel through a leased line network being routed by mini-computers like cars on a railroad.

The use of such private networks will reduce the need to overlay separate systems as often happens today, and so will lower the overall data communications bill of a corporation.

3. *Combining the Transmissions of Separate Organizations*

A higher utilization of shared facilities may be achieved if several different organizations combine in the sharing. A network of leased lines and private switching facilities could serve an association of several corporations at a lower cost per user than if it were designed for one corporation. Unfortunately, this is not permitted by the regulations controlling the use of telecommunications in most countries. In the United States, it is permitted only for certain special cases.

4. *Line Broking*

An entrepreneur might operate a "line broking" business in which he leases lines, voice or wideband, attaches equipment to them that permits them to be shared, and then sells the subchannels so derived to terminal users. The line broker may be one of the major users with capacity to spare. This would be a very valuable service in data transmission because, again, it would greatly lessen the wastage and hence the cost to the user. Unfortunately, line broking is also prohibited in most countries.

5. *Public Data Networks*

In the example at the start of the chapter, there were no other users available with whom to share the line across the Atlantic; thus it had to be used in a wasteful manner. Such is often the case today, especially when public telephone lines are used for data. To avoid this inefficiency, a switched public network, designed for data transmission is needed. Such a network would make use of high-capacity trunks that carry many millions of bits per second. The Telex and TWX networks are examples of switched public networks, but these can only operate at low speeds. It is desirable that a switched public data network should operate at speeds ranging from about 200 bits per second to 48,000 or higher.

FUTURE EVOLUTION OF COMPUTER NETWORKS

The need for communication line multiplexing is a basic factor in achieving economic usage of remote computers. A *very* high degree of multiplexing is necessary to take advantage of the digital transmission media that today's technology makes possible. We can now build microwave links and coaxial cables that carry a billion bits per second. Consequently, in the years ahead we are likely to see an evolution of data transmission networks passing through the following stages:

1. *Private networks in which terminals are linked to one system* designed for one set of functions—for example, today's airline, banking and data entry systems and time-sharing systems using one computer center.
2. *Private networks in which more than one system are interlinked, within one corporation.* For example, the interlinking of separate time-sharing facilities or data-base systems; the interconnection of computer centers.
3. *Private networks in which systems in different organizations share data network facilities.* For example, interbank systems or interairline systems in which network facilities are shared and, in some cases,

computing services are shared; the interconnection of time-sharing systems and computer centers in different organizations.

4. *Public data transmission networks that cover a portion of a country only.* Unlike the telephone network, they will be built to take advantage of digital technology and multiplexing so that the user pays a fee proportional to his data rate. For cost reasons, they will probably only cover a portion of a country at first—the profitable portion with a high density of users. Examples: The Datran system in the U.S.A., Germany's EDS system, Britain's proposed data network.
5. *Nationwide public networks.* It may be two decades before public data networks become as ubiquitous as the telephone network. Eventually terminals in most homes and offices will be interlinked.
6. *International public data networks.* With the rapidly dropping cost of international links, it seems probable that international networks "cream-skimming" the most profitable routes may exist before national networks give full coverage to less-profitable areas. The problems we are now creating with international incompatibility will have to be solved with interface computers.

The further we progress toward nationwide data networks, the less wastage of bandwidth there will be. It has been estimated that there will be 2.5 million terminals in use in the United States by 1980. If all of them used their communication lines as inefficiently as the example at the start of this chapter, the wastage of the available lines would be appalling. It is almost certain that sufficient communication lines to support this usage could not be provided. Private networks, as discussed in the next chapter, would be somewhat less wasteful. However, many separate private networks overlaid geographically require a much greater number of communication lines than a public data network with digital microwave links or coaxial cables and a very high level of multiplexing. Such a data network is going to become an extremely important national resource.

11 NETWORK STRUCTURES

With increasing numbers of terminals being used with one computer system, the need to devise means of lowering the overall cost of the network continues to grow.

Many terminals transmit and receive data at a relatively slow rate. In a large network, the terminals may be connected to low-speed lines which are tributaries of higher-capacity lines. In some cases, these higher-speed lines may themselves be tributaries of a line of still higher bandwidth, like streams flowing into bigger streams and eventually joining a river that rolls down to the sea.

USE OF A MULTIPLEXER

Many of the smaller streams may be brought together at a point to form a larger stream. This step can be accomplished with a device called a multiplexer, shown in Figure 11.1.

The higher-speed line in this diagram is often a leased full-duplex voice line, operating at, say, 4800 bits per second. The lower-speed lines may be half duplex, operating at 75 or 150 bits per second. They may link teletype machines or typewriterlike terminals to the multiplexer. The multiplexer scans the low-speed lines and retransmits the results of each scan down the high-speed line to a computer. It sends a block of data on the high-speed line which consists of either a bit from each low-speed line, or a character from each low-speed line (including NUL characters when a terminal is not sending). At the same time, it receives data from the computer which are to be distributed to the terminals by a converse process. One typical machine can handle 24 lines of 150 bits per second in this way.

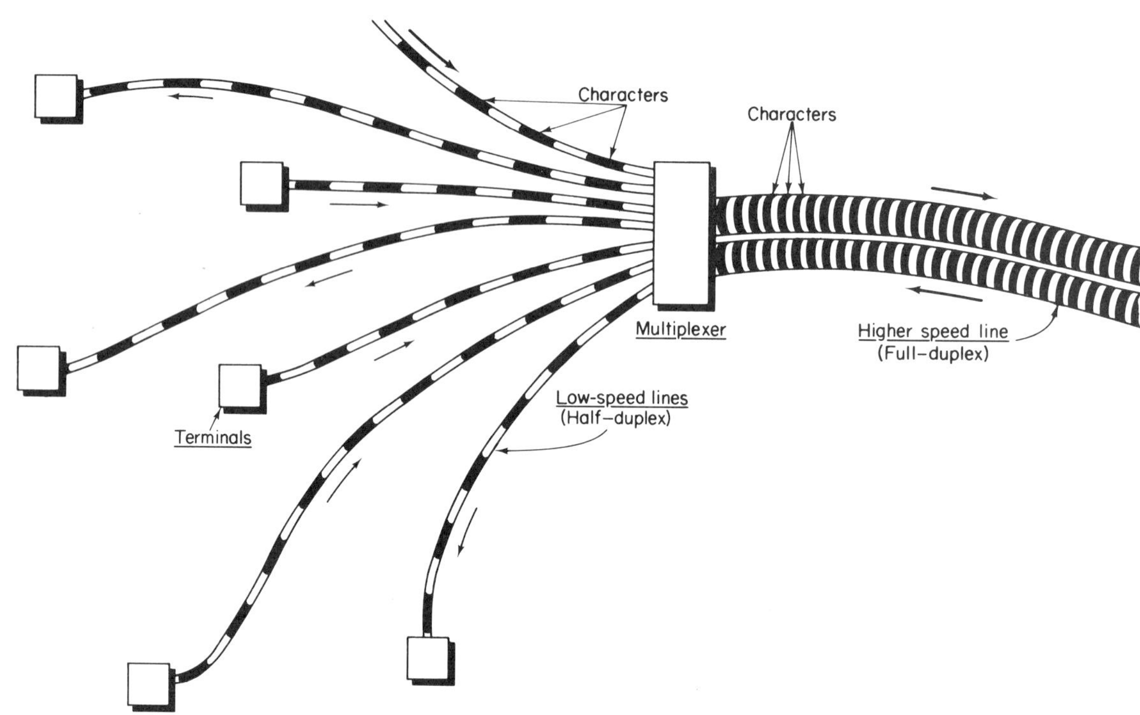

Figure 11.1. Time-division multiplexing.

An alternative way for a multiplexer to work is to assign a separate *frequency* range on the high-speed line to each of the low-speed lines. In this way the bandwidth is sliced up into smaller bandwidths (Figure 11.2).

Whichever way it operates, the signals from a number of lower-speed lines are combined to travel over a higher-speed line, with the data structure essentially unchanged.

TAKING ADVANTAGE OF A START-STOP OPERATOR

With simple multiplexing, *all* the terminals can transmit *all* the time. Most terminals do not need to do so. As we illustrated in Figure 9.8, when an operator is carrying on a dialogue with a terminal, the data flow in bursts, often small compared with the amount of data that would be sent if the line transmitted continuously. Furthermore, when an operator uses a keyboard, the resulting character rate is usually far less than that of the

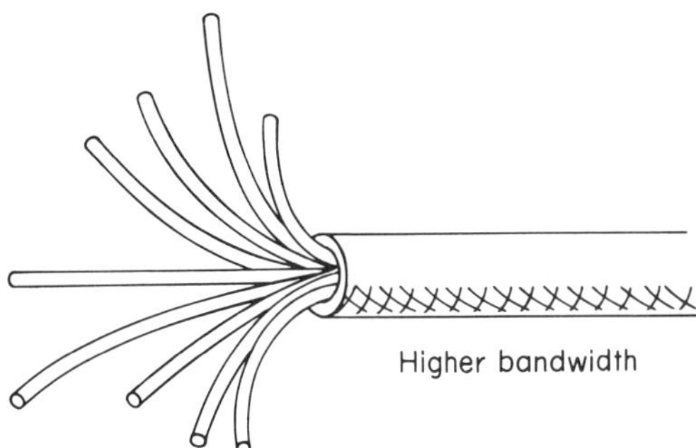

Figure 11.2. Frequency-division multiplexing.

line. An input rate of three characters per second would be faster than that of most terminal operators, but this is a fraction of the 150 bits per second that is possible on many subvoice-grade lines (200 bits per second in many countries).

If the low character rate is taken advantage of, the terminals in Figure 11.1 might be handled without the need of a higher-speed line. This situation is illustrated in Figure 11.3. To be satisfactory, the device in this diagram, referred to as a concentrator, needs to have storage that is flexibly allocated to the different terminals. Storage must be sufficient to absorb the temporary overload that will occur when, fortuitously, all the operators happen to pound away at their keyboards simultaneously and for a while generate data at more than 150 bits per second. The concentrator in Figure 11.3 collects the characters arriving from the terminals and sends them onward in blocks or messages. The characters on the lines connected to terminals are widely scattered, but on the line from the concentrator to the computer they are packed together. It is rather like people driving to a bus station, one person per car, and all getting into a bus. Both the cars and the bus use the same road facilities.

A block or message from the concentrator must indicate which terminal its characters come from. Conversely, blocks traveling to the concentrator indicate which terminal they are bound for.

Figure 11.3 shows half-duplex lines to the terminals but a full-duplex line from the concentrator to the computer. A half-duplex line could have been used from the concentrator to the computer. Although simultaneous two-way transmission between the terminal and computer would give little improvement, two-way transmission between the concentrator and computer increases the capacity of this line substantially and would make it possible to increase the number of terminals connected to the concentrator.

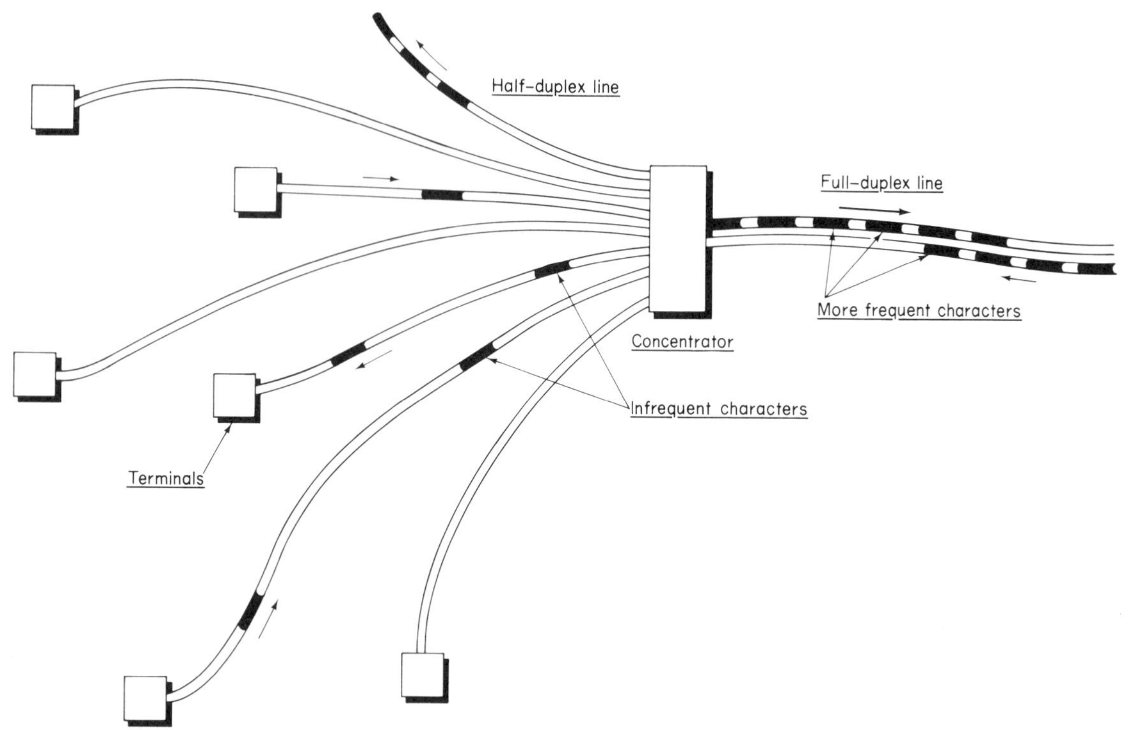

Figure 11.3. A concentrator taking advantage of the infrequent character transmission from the terminals. Several low-speed lines are connected to a full-duplex line of the same speed.

TAKING RISKS

The configuration in Figure 11.3 avoids the cost of the high-speed line but is taking a risk that input data will occasionally have to be reentered or that a keyboard will have to be automatically locked because of a temporary overload when all the terminal operators enter data at once.

In the design of transmission networks, there are other such risks that the system planners sometimes take in order to lower the cost of the transmission facility. In some cases, the design uses line switching and runs the risk of giving thc terminal operator a "busy" signal, blocking his ability to transmit. In other cases, the design will result in lengthened response times when an occasional overload occurs.

The designer should plan his risk taking to minimize the frustration it causes for the terminal operators. A bad form of risk would be that of losing data. Equally frustrating is to be cut off in the middle of a terminal dialogue, especially if it is one in which it will be difficult to pick up the threads. Another bad design would be one in which the operator, when frustrated by busy signals or long response times, can take some action that, although it might relieve his feelings, will make the network condition still worse.

The designer must evaluate the properties of potential networks by doing appropriate probability calculations [1].

MULTIDROP LINES

A good way to take advantage of sporadic terminal activity is to connect several terminals to a single line. Figure 11.4 illustrates this process, showing a *synchronous* message going from the computer to terminal 2. It could also be an asynchronous (start-stop) message, as in Figure 7.4. A line with several devices, or *drop points*, is called a *multidrop* line.

There can only be one message traveling at once on a multidrop line; otherwise the line conditions representing bits become jumbled and the bits would not be interpreted correctly. The reader should recall when looking at Figure 11.4 that each pulse travels from one end of the line to another in a flash—at the speed of electrical propagation, which for some transmission media is close to the speed of light. Although there may be only 75 bits per second transmitted, each bit makes its presence felt over the entire length of the line almost simultaneously. The "fire-hose" style of diagram in this chapter fails to convey this almost instantaneous propagation. Nevertheless, the reader can no doubt reflect that if our espionage fireman had several agents connected to the firehose over which data were being sent, then only one of them could be transmitting at once.

The fireman, in fact, must have some way of maintaining discipline over his fellow agents. They must have a carefully worked out procedure for who transmits when. Discipline is needed on multidrop communication lines for the same reason and will be discussed in the next chapter.

When the fireman sends a pulse down a multidrop fire hose, it will be detected by all the agents connected to that hose. The same is true for the bits that are shown traveling down the line in Figure 11.4. Most messages that are sent, however, are intended for only one receiving device. The message in Figure 11.4 is for terminal 2. The other terminals must ignore it. Each message sent down the line, therefore, has an address in it in a fixed position. One character is sufficient for the address, and it is often the first character. The control unit of each terminal examines this character. If it is not the address

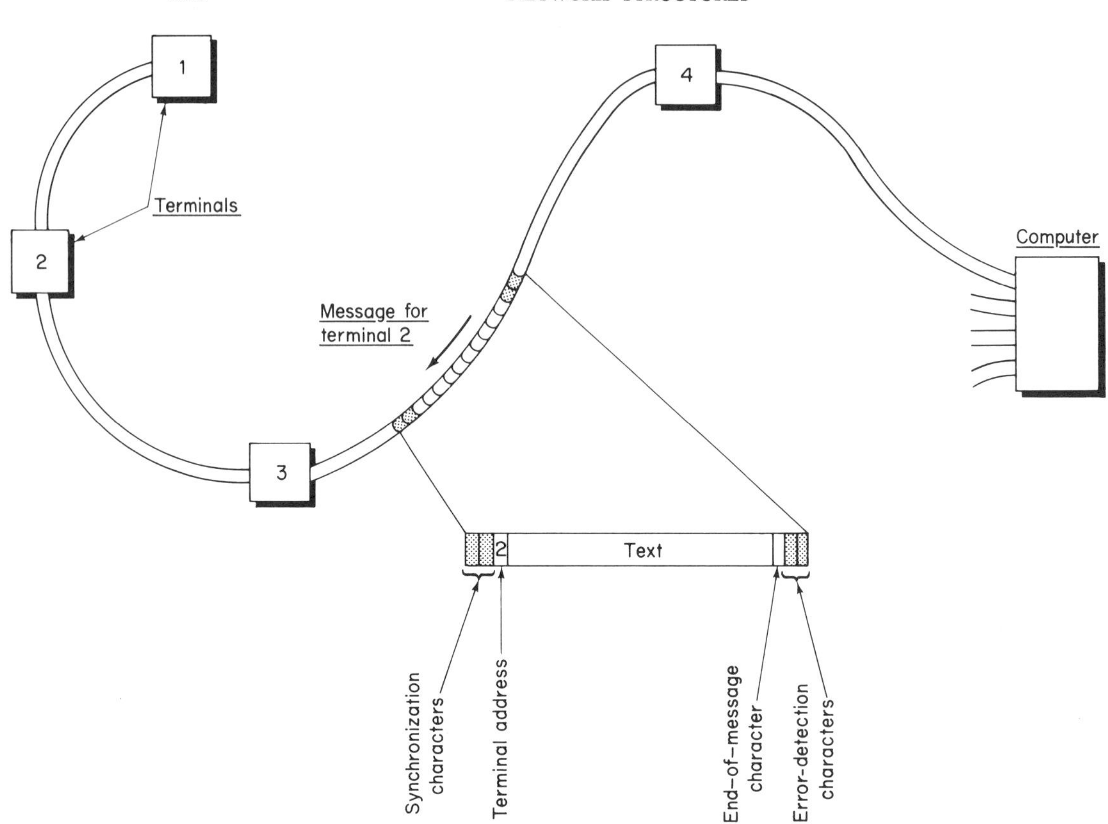

Figure 11.4. A multidrop line. Every message sent from the computer reaches every terminal. The only terminal to accept it is the one whose address is at the front of the message. Disciplined control of the line must ensure that only one device can be transmitting at a time; otherwise an undecipherable jumble of characters would result.

of that terminal, the message will be ignored. In some systems there is a "broadcast" address, in addition to individual addresses, meaning that the message should be received by *all* devices on the line.

The *risk* involved with multidrop lines is one of degraded response time. If terminal 3 wants to transmit to the computer, it has to wait until the transmission shown in Figure 11.4 is completed. It may have to wait longer, taking its turn after other transmissions. Occasionally, and by chance, all the devices will be ready to transmit at once; one will have to wait until all the others have finished. The response time, then, will be uneven. It will occasionally be lengthened by the waiting while other devices transmit.

If every terminal has a buffer and transmits data from its buffer, the wait may never be very long. When a terminal does not have a buffer, the time it takes to send a message depends on the rate at which the operator types it. The risk is therefore much greater when the terminals are unbuffered. A slow or disruptive operator could occasionally cause the other operators to be held up. Some multidrop systems have been designed so that if an operator pauses for more than a given period, say 5 seconds, the opportunity is given to other devices to break in. Ingenious operators (and most of them learn such tricks) discovered that they could avoid losing the line in this way by idly fingering the *Shift* key, causing a sequence of Shift characters to be sent and thus never leaving a 5-second gap between characters.

For a real-time system in which a fast response time is necessary, the risk of unbuffered terminals on a multidrop line is generally not worth running, although buffered multidropped terminals are often used.

CONCENTRATORS ON MULTIDROP LINES

A variety of combinations of the preceding techniques are possible. A channel derived by dividing up the bandwidth of a larger channel (Figure 11.2) could itself be used for multidrop operation. A *multiplexed* line could link into a concentrator. A concentrator could occupy one channel of a multiplexed line. *Multidrop* lines could link into a concentrator, or concentrators themselves could be multidropped. The latter situation is shown in Figure 11.5.

Whereas a multiplexer may convert several low-speed transmissions into one high-speed transmission, a concentrator may convert several inefficient transmissions into one efficient one. It may do both—changing both the speed and the transmission mode. Synchronous transmission gives more efficient use of a line than start-stop transmission. One function of a concentrator can be to convert start-stop transmission into synchronous transmission.

The concentrators in Figure 11.5 change the speed and the transmission mode, *and* operate on multidrop lines. The use of multidropping may introduce less delay than on Figure 11.4 because high-speed lines are multidropped rather than low speed.

LINE SWITCHING

Many networks employ a line-switching mechanism. The terminals are then not permanently connected to the computer but are on a line that must be switched to it. The operator may dial the computer number to accomplish the switching; or if the connection is always to the same computer, the switching may be automatic.

In a leased-line network, the total line mileage can be cut greatly by using line switching. Figure 11.6 shows a switching mechanism. Messages from the

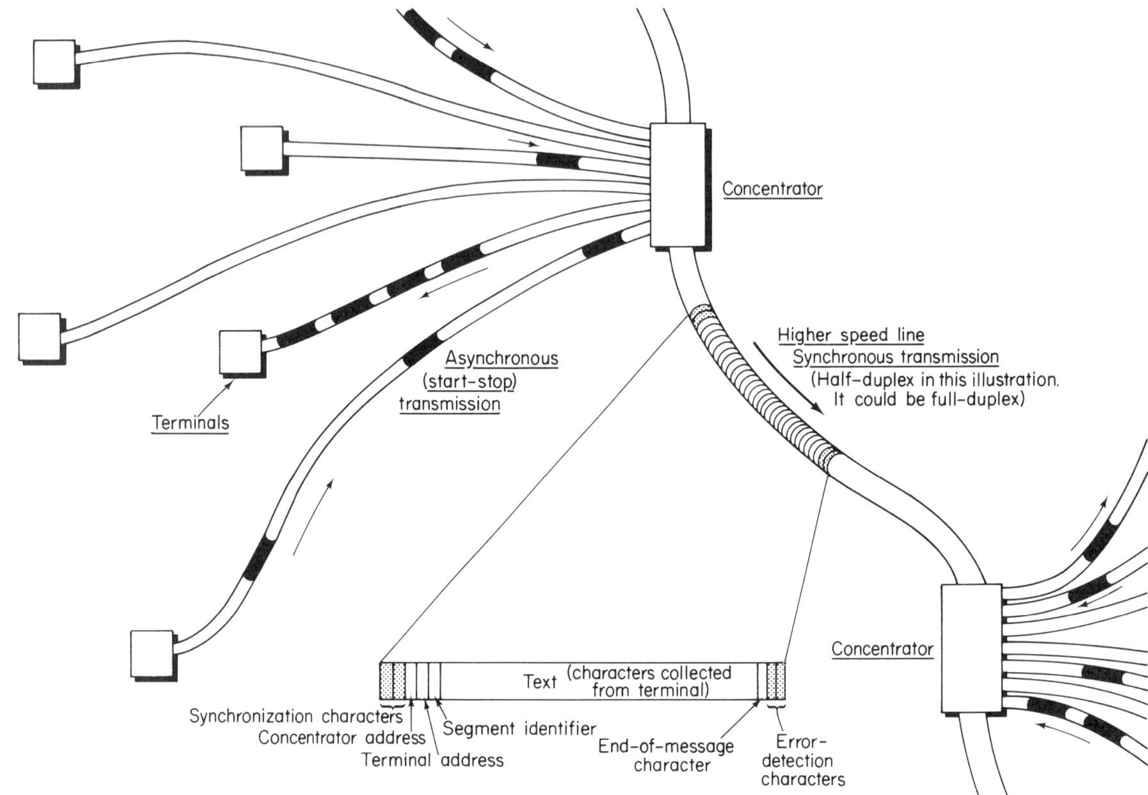

Figure 11.5. Concentrators on multidrop lines.

computer to a terminal go through this switch. The messages may be preceded by addressing characters, which cause the line from the computer to be physically connected to the line to the appropriate terminal. Similarly, when the terminal sends a message to the computer, the appropriate connection must be established. There may be more than one line from the exchange to the computer, and the switching mechanism must be able to *hunt* for a free one.

The switch mechanism may be a privately owned switching device; it may be a private branch exchange leased from the common carrier; it may be a mechanism installed in the local central office (telephone exchange). Most switching equipments today are relatively slow electromechanical mechanisms [2]. In the future, fast electronic switching with no moving parts will become more economical [3].

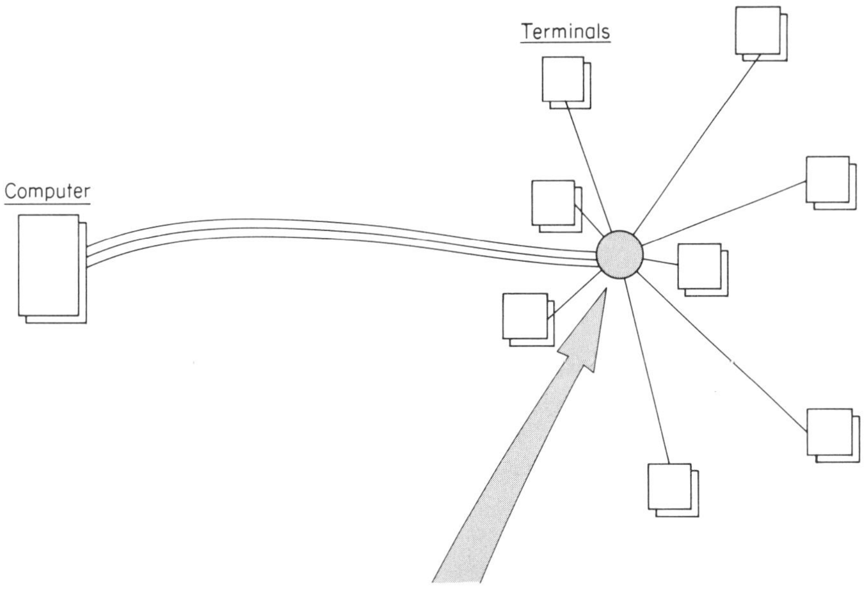

Figure 11.6. Private exchange or switching mechanism. The switch is often an electromechanical device, as in the diagrams below.

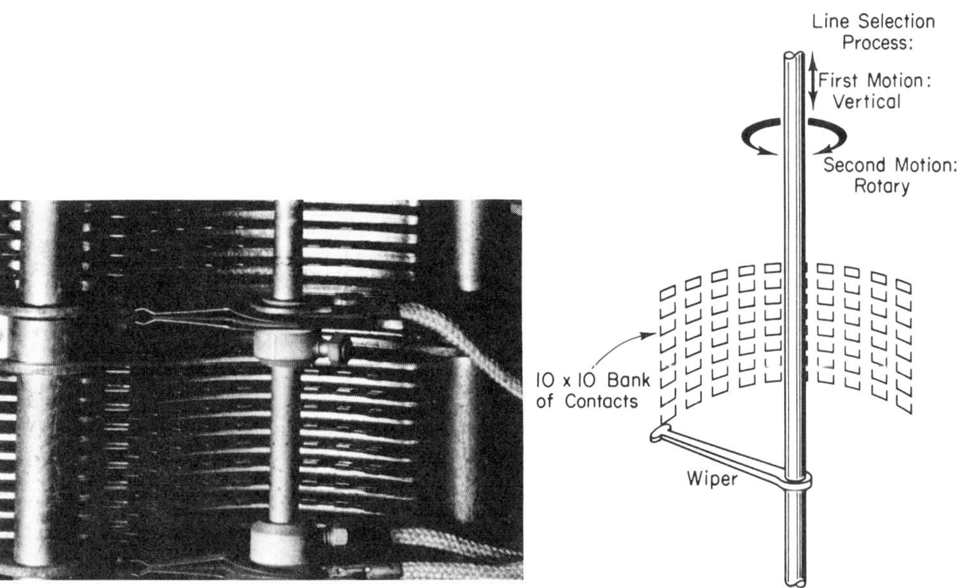

The *risk* with line-switching networks is that the path to the computer may be already occupied. It will not be possible to set up a path when the operator requests it. The operator will obtain a "busy" signal, as on the telephone, or its equivalent.

The probability of failing to make a connection will depend on the number of lines from the switch to the computer and on the number of terminals and their usage. The designer must calculate the probability of the terminal operator not being able to make a connection and being kept waiting for a long time [4].

LINE SWITCHING VERSUS MESSAGE SWITCHING

There are two fundamentally different types of switching. The physical transmission path may be switched as in Figure 11.6, or the messages may be received at a switching center, stored, and then routed onward to the requisite destination. The former is referred to as *line switching* and the latter as *message switching*.

For years message-switching systems have been used to relay cables between many locations. Originally message-switching centers were manually operated. Operators took the messages received and forwarded them to destinations shown in the message headers. Many of these manual centers are still in operation. Today, however, computers are being used for message switching; and in addition to relaying cables, they form part of data-processing networks.

Message-switching centers can give better line utilization than if the same network was operated with line switching. Although the switching cost is higher, the total line cost is lower. Message switching therefore tends to be economical on a large network with long lines and many terminals.

A message-switching computer carries on two functions continuously: it receives messages and stores them in its file, and it takes the messages from its file and sends them to the requisite destination [5]. It often uses multidrop lines with many terminals. When usage becomes heavy, a message-switching center does not fail to make a connection, as does a line-switching device (Figure 11.6). Instead it slows down its delivery operation. Message switching can form a "nonblocking" network that never fails to accept traffic unless there is a breakdown. The *risk* with message switching is not of "busy" signals, as with line switching, but of a lengthened delivery time.

PACKET-SWITCHING NETWORKS

Message-switching systems have traditionally been used to relay messages that do not need a very fast delivery time. Recently some systems have been built, however, that operate somewhat like message-switching systems and that relay their messages in a few milliseconds. These have been referred to as packet-switching systems.

In such systems, the terminals or data-processing computers are linked to the network via small *interface computers.* The interface computer forms the data that are to be sent into "packets," often of fixed length with a fixed format header. Wideband communication links interconnect the computers, the network being designed so that if one of the lines fails, there is always an alternate line to the destination computer. The packet-switching computers form the nodes of a multipath network. They determine the routing for the "packets," and the packets are passed very quickly from one node to another. The packet is never put in secondary storage as in a message-switching system.

The ARPA (Advanced Research Projects Agency) packet-switching network linking computers in several American universities and research centers uses 50,000-bit-per-second lines, and the packets normally traverse the network in less than 100 milliseconds. Packet switching can thus be used for fast response-time systems. The British Post Office intends to incorporate packet switching (as well as line switching) in its eventual public data network.

SUMMARY OF RISKS

With different types of network structures, we thus run different types of risks. Table 11.1 summarizes them.

In addition to those listed, there is one other important risk. The communication links or equipment might fail. If high *availability* is important, the network may be built with alternate transmission paths. Sometimes when a leased line fails, a public connection may be dialed. Sometimes more than one leased path interconnects vital locations. In the ARPA network, if any wideband link fails, an alternate route is selected by the packet-routing nodal computers.

THE LOCATION OF MODEMS

For simplicity, the diagrams in this chapter have not shown the modems. If the lines were digital in structure, modems would not be needed. Today they are almost always analog, and so modems would be used to connect each data-processing machine, including digital multiplexers and concentrators, to lines.

Figure 11.7 shows the typical positioning of modems in a network.

Note that the multidrop line on the left is shown with a fork in it. This situation is quite normal. An electrical pulse originating at one of the devices on the line would travel almost instantaneously down both branches of the fork. The multiplexers are shown as being digital machines. If, however, they

Table 11.1. TABLE OF RISKS

Network Device	The Risk That Occurs During Periods of Overload:
Simple multiplexer (Figs. 11.1 and 11.2)	No risk related to overload. All devices are permanently connected.
Hold-and-forward concentrator (Figs. 11.3 and 11.5)	Terminals may be cut off or responses delayed. (In a badly designed device, messages may be occasionally lost if a transmission cannot be cut off.)
Exchange or line-switching device (Fig. 11.6)	It may not be possible to establish a connection; a "busy" signal will be received.
Multidrop line (Fig. 11.4)	Response time will be lengthened.
Message-switching system	Delivery time will be lengthened.

Other Risks

1. Risk that a transmission error will be undetected.
2. Risk that transmission equipment will fail.
3. Risk that a leased line will fail. (A line outage is of little concern on a public network because dialing will select an alternate path.)

divided the bandwidth up by frequency (Figure 11.2), they would not require the modems shown connected to them, for there would be no analog-digital conversion. The modems would simply be used at the terminals and at the computer.

SHORT, MEDIUM, AND LONG LINES If the communication lines are long and therefore expensive, it will be worthwhile using elaborate techniques to reduce the number of lines needed. The larger real-time communication networks of today would have been unthinkably expensive without the concentrators, multiplexers, or other devices that they use, and without the sometimes elaborate line-control procedures.

On the other hand, if the lines are very short or inexpensive, the emphasis will be on lessening terminal and equipment cost rather than line cost. Where the lines are very short—for example, all in one plant, one office block, or one campus—then their cost is of little significance in the design, and we

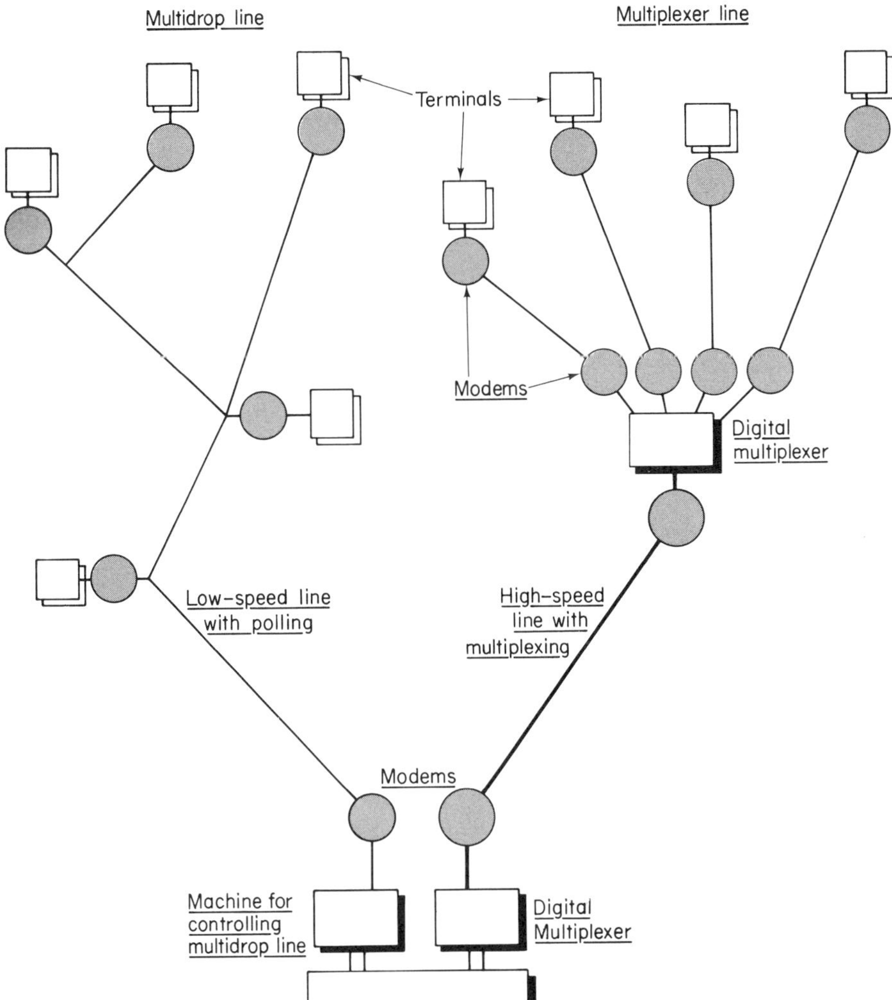

Figure 11.7. The locations of modems. The network structure diagrams in this chapter have not shown the modems. On analog lines, they will be needed in the positions shown.

often find their bandwidth being used quite wantonly. There may be multiple wires to each terminal and transmission that is parallel by bit, as in Figure 7.1. There may be a high-speed loop, as in Figure 7.8, with the majority of bits being used to simplify the control logic rather than to carry data.

Where a system has a very large number of terminals, the terminal cost may dominate the design. For this reason, many systems have used start-stop transmission from the terminals, and no buffering in the terminals. This has made possible an inexpensive terminal design, as with telegraph machines. Buffering and conversion to synchronous transmission are then sometimes done at a concentrator, as in Figure 11.5.

The cost of placing logic and storage (such as buffers) in the peripheral parts of the network is dropping rapidly, largely because of large-scale integration circuits. As this happens, there will be an increasing number of systems that use multiplexers, concentrators, and other remote mechanisms for lowering the overall network cost.

The trade-off calculations necessary to design a minimum-cost network are becoming very elaborate [6].

12 DISCIPLINE AND THE TRANSMISSION CONTROL UNIT

Discipline is needed with many forms of transmission to ensure that two devices do not transmit at the same time on the same channel. It is needed with people, too. The policemen and soldiers in television thrillers all say words such as "over" and "out" when talking over their radio networks. In data transmission, the machines say "over" and "out" by sending special characters —for example, end-of-message and end-of-transmission characters.

One popular scheme uses seven special characters with the meanings indicated in Table 12.1.

Table 12.1

Character	Meaning	Bit Pattern
SOH	Start of header	1000000
STX	Start of text	0100000
ETX	End of text	1100000
EDT	End of transmission	0010000
ACK	Acknowledge affirmatively	0110000
NAK	Acknowledge negatively	1010100
CAN	Cancel	0001100

After each text message is received, an acknowledgement is sent saying whether it was received correctly or not. ACK means that it was received correctly and NAK means that it was not. A cancel (CAN) character placed in the message means that it should be ignored. The transmitting machine found a reason to cancel the message when it was too late to stop its transmission. The control of errors is thus part of the line-discipline procedures.

Some machines use more control characters than the preceding ones. As can be seen in Figure 6.6, the United States standard ASCII code has provision for many control characters.

When two or more machines are attached to a communication line, it is generally the case that one machine will be the controlling station. If there are terminals attached to a computer, the computer will normally be in control. When terminals communicate between themselves, one is usually designated as the master station.

MULTIDROP LINES

Maintaining discipline is more complex on multidrop lines than on point-to-point lines.

When several devices all share a communication line, only one device can transmit at once, although several or all points can receive the same information. Each device therefore has an address, usually of one character. It must have the ability to recognize a message sent to that address. A line may, for example, have 26 terminals with addresses A to Z. A message for terminal A may be preceded by a message saying "terminal A, are you ready to receive." or it may have the address A in the header of the message itself.

Sometimes the line discipline may permit the sending of a message to more than one device. Suppose that the computer sends down the line a message that is to be displayed by terminals A, G, and H. The message is preceded by these three addresses, and each terminal has circuitry that scans for its own address. Terminals A, G, and H recognize their addresses and display the message simultaneously. The other terminals do not recognize their address and so ignore the message. The network may also have a "broadcast" code that causes all terminals on a line to display those messages preceded by it.

POLLING

For transmission in the other direction, several terminals may wish to transmit at the same time. Only one can do so, and the others must wait their turn. To organize this process, the line will normally be polled. A *polling message* is sent down the line to a terminal saying, "Terminal X, have you anything to transmit? If so, go ahead." If terminal X has nothing to send, a negative reply will be received and the next polling message will be sent, "Terminal Y, have you anything to transmit? If so, go ahead."

As a rule, the computer organizes the polling. The computer may have in its storage a polling list telling the programs the sequence in which to poll the terminals. The polling list and its use determine the priorities with which terminals are scanned. Certain important terminals may have their address more than once on the polling list, so that they are polled twice as frequently as the others. Certain terminals may always be polled before the others.

Some computers can receive and transmit on all their communication lines at the same time. Other hardware is more restricted, and only a certain number of lines may be in use at once. In the latter case, the polling concept must be extended beyond one line, and the polling list will relate to several lines.

One problem with this form of polling is that where high-speed transmission is used—say 4800 bits per second on a voice line—the line turnaround time is long compared with the transmission time. "Roll-call" polling may necessitate many line turnarounds (reversals in the transmission direction) before a particular terminal is polled, which can degrade the response time. The problem can be overcome by *hub polling*, in which each terminal passes the polling messages on to its neighbour, and by line disciplines involving continuous transmission on a loop [1].

THE TRANSMISSION CONTROL UNIT

At the computer center, the reception and transmission of bits must be controlled and the line-discipline operations must be carried out.

The digital information is transmitted on the communication lines a "bit" at a time. The messages to be transmitted must therefore be broken into bits, and these bits must be sent at the speed of the line. Similarly, on receiving a message, bits are assembled one at a time into characters and the characters are assembled into messages. Both the characters and messages must be error checked and the errors corrected, if possible. Suitable control signals must be generated for operating the distant terminals at the correct times.

Depending on the nature of the hardware in use, this may be done in the following ways:

1. Entirely by programming.
2. Entirely by electronic circuitry.
3. Half and half—for example, with the electronics assembling bits into characters and the programs assembling characters into messages.

At the computer center, the communication lines may terminate in one of four different ways.

1. They may go directly into the main computer, so that bits arriving are stored directly in the main memory. This approach may result in many interruptions and "cycle stealing" from the main operations. It is often restricted to systems with low-volume transmission requirements.
2. They may terminate in a *transmission control unit* or *line control unit*, which feeds characters to the main computer and accepts data characters and control characters from it. The main programs will then be interrupted to assemble characters into messages or to feed characters to the line-control equipment.
3. The transmission control unit may pass blocks or words, rather than characters, to the main computer and may receive blocks or words from it. There is then less time taken from the main computer programs.
4. The transmission control unit may itself be a stored program computer, which feeds complete, checked-out, edited messages to the computer and receives complete messages for transmission. This transmission control computer would assume responsibility for maintaining discipline on the lines and would leave the main machine free to do the processing.

The choice among these approaches may be indicated to some extent by the number of communication lines that must be handled. If the system has only one communication line, this may go straight into the computer or into a device with a buffer storage, which automatically assembles or disassembles a complete message. If, however, there are 50 or 100 communication lines, these may best be terminated in a separate line-control computer.

The value of a stored-program computer in handling communication lines lies in its adaptability. It must handle messages of any length; therefore dynamic memory allocation is desirable. It may control differing numbers of lines, with different devices. It will often have to change the sequence in which it polls terminals, as terminals are shut down and opened up. It may have to dial out to terminals and accept calls dialed in. Diagnostic programs will be run in it to detect network faults and help in terminal check-out.

The line-control computer may have an instruction set that is different from conventional computers and designed for handling communication lines. It may have the facility to log messages on its own random-access file or tape unit. It may send English-language messages to the terminal operators as part of its control procedures. Because it is a programmable unit, its procedures may be modified as circumstances demand.

CODE CONVERSION

An additional function of the transmission control unit may be code conversion. The transmission line often uses a different means for encoding characters to the main computer. Seven-bit ASCII and five-bit Baudot codes (Figures 6.6 and 6.1), for example, are standard on communication lines, whereas eight-bit EBCDIC code is commonly used in the main computer. Either the transmission control machine or the main computer must translate the code. This step is now often done by hardware or microprogramming.

THE DEVICES IN A NETWORK

Figure 12.1 illustrates the devices used on a teleprocessing network.

The functions of the transmission control unit are shown; and if it carries out all of these functions, it is probably a stored-program computer. It could, however, be a microprogrammed or nonprogrammed device, and some of the functions shown could then be in the main computer.

In this diagram the transmission control unit handles two types of lines. Such a device could handle several different types of lines. On the other hand, where a nonprogrammed device that merely assembles and checks characters is used, there may be different such devices for different line types or groups of lines.

The line on the left-hand side of the diagram is a multidrop line. The line-control unit is shown polling it and maintaining the multidrop discipline. The line on the right-hand side uses multiplexing (as in Figure 11.1). The line-control unit is shown doing the multiplexing (and demultiplexing) at the computer end of the line. In some systems a separate multiplexer unit does this operation (as in Figure 11.7).

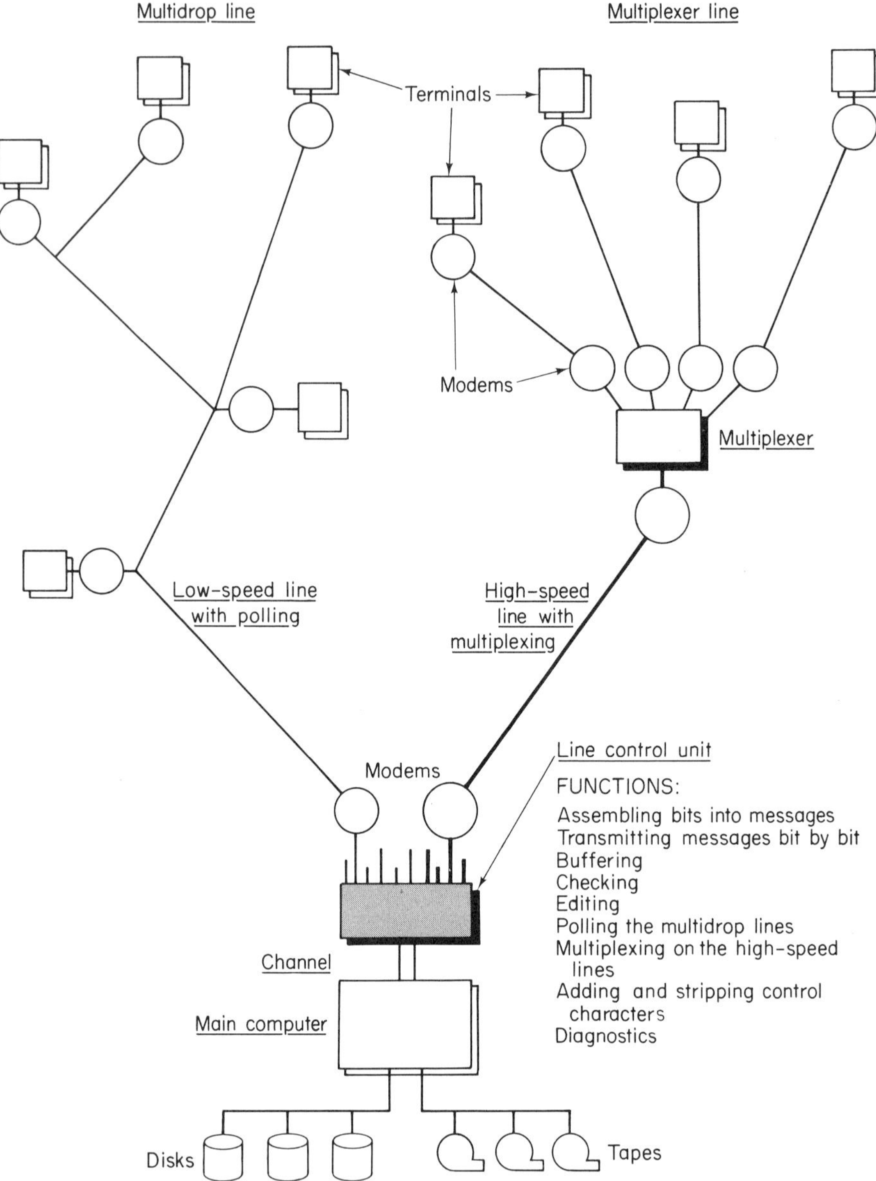

Figure 12.1. Functions of the transmission control unit. Some of these functions may be in the main computer.

Figure 12.2. Transmission Control Units

1. A Simple Non-Programmable Transmission Control Unit

The IBM 2701 Data Adapter Unit that attaches up to four half-duplex lines to an IBM 360 or 370. The lines can be START/STOP lines of speeds up to 600 bits per second, or synchronous lines of speeds up to 230,400 bits per second. The 2701 sends the data to the computer, and receives it from the computer, one character at a time.

2. A Stored-Program Computer as a Transmission Control Unit

The IBM 3705 Communications Controller attaches up to 352 low-speed lines, or smaller numbers of higher-speed lines, to an IBM 360 or 370. It is a stored-program computer with a core size ranging from 16,000 to 240,000 bytes. It relieves the main computer of teleprocessing functions such as line control, polling, address, code translation, and error recovery, and can deliver complete edited messages to the main computer. It continues to control the communication lines if the main computer fails. Being a stored-program machine, it has the flexibility to control all types of transmission.

13 TERMINAL CONSIDERATIONS

Chapter 1 listed the common types of terminals. Now that we have explained the transmission mechanics, we will discuss terminals in more detail.

Two factors are involved in terminal design: those relating to the terminal's interface with the telecommunication network and those relating to its interface with people, or means of handling data. These two aspects link together in the overall consideration of the logic or "intelligence" that is built into the terminal, and hence, to a large degree, its cost.

In the case of a man-computer dialogue system, terminal features will form an essential element of the dialogue. A poorly planned terminal configuration can have a serious effect on the efficient operation of a system. It can slow down the system, introduce errors, lessen the usefulness of the system, and significantly add to its overall cost. At worst, it can reduce a normally calm individual to a state in which he wants to put his foot through the screen.

From the average user's point of view, the terminal is very important. It is the only component of the system he sees. He forms his impression of the system entirely from the terminal.

It is not the purpose of this book to discuss the man-machine interface [1]. We will merely say that for efficient dialogue the terminal must be as unconfusing as possible and must give a suitably fast response time or output speed [2]. For most uses, it should be operable with one hand. (Many terminals need two hands because of shift and alternate coding keys.) The screen on a visual display should be large enough for the dialogue in question. There should be an easy means of responding to the screen contents. The labels on the keyboard form an important element of the overall dialogue structure. Given the right terminal, the dialogue structure becomes all-important to the operator, but the dialogue design may be regarded as a combination of terminal hardware design and design of the data content.

Two aspects that are sometimes neglected when the terminal is being chosen are security and the question of what happens when the communication line or computer fails. Terminal features are sometimes important in the overall plans for maintaining security and for recovery from failures.

THE INTERFACE WITH THE COMMUNICATION LINE

In a teleprocessing system there are three components which are incompatible in speed: the computer, the terminal, and the communication line linking them. They are usually incompatible in other ways as well. The computer and most terminals are digital devices designed to deal with bits, or square-edged pulse trains; but most communication lines are *analog* in operation, designed to transmit a continuous range of frequencies. The coding of characters used in most terminals is different to that in the computers they are connected to. The error rates encountered on most communication lines are higher than those generally acceptable in computers.

The modem solves the problem of analog-digital incompatibility. Error detection and retransmission solve the problem of error rates. The problem of code conversion is left to the transmission control unit or to the computer. But for the question of speed, the terminal is on its own.

INCOMPATIBILITIES IN SPEED

Since its earliest days the computer has had the problem that it operates much faster than its input and output units. Being an expensive machine, it cannot afford to wait for them. The problem was solved by using *buffers.* The computer dumps a quantity of output into a storage unit, or buffer, at full speed, and then the contents of this unit are printed slowly, at the speed of the printing mechanism. When the information has been printed, the computer again refills the buffer at its own speed. The converse takes place for input.

A communication line is handled in a similar way. The characters from the computer will normally be transmitted one bit at a time at normal line speed from a buffer that is refilled periodically. It is somewhat like eating pudding with a large spoon. A person chews and swallows pudding at a fairly slow rate (an English pudding), so he uses the spoon as a buffer and refills it quickly at appropriate intervals (Figure 13.1).

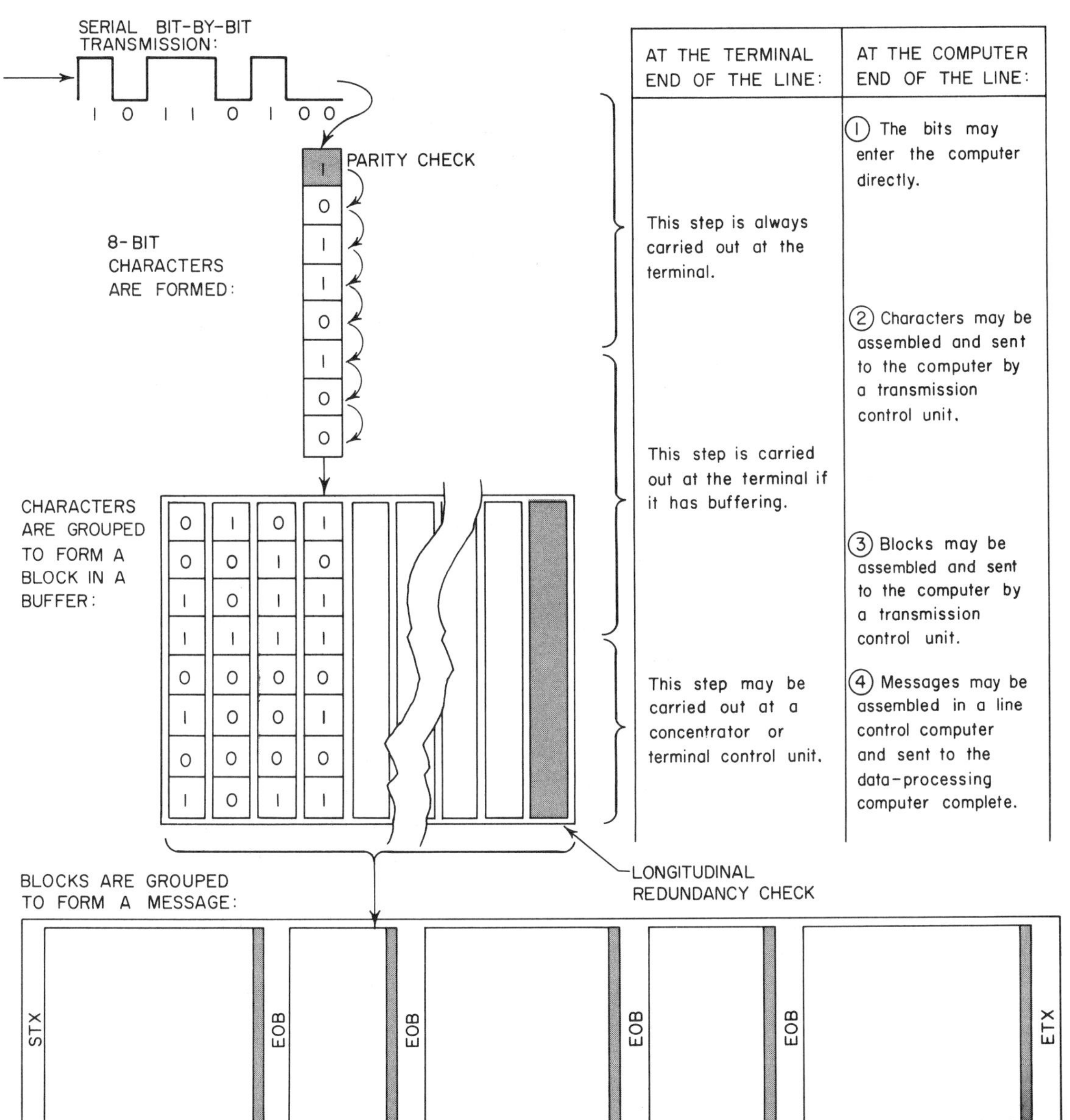

Figure 13.1. Bits, characters, blocks, and messages in data transmission. (A converse process takes place in output.)

TERMINAL BUFFERING

The terminal may also have a buffer. Suppose that a voice line transmits at 4800 bits per second from keyboard terminals. The operators cannot type information into the terminal at this speed; at best, they may type about three characters (21 bits) per second, but there will be lengthy pauses for thought and other activity. In this case, their keying might fill up a buffer of, say, 100 characters, the contents of which would be transmitted, in turn, over the line in one burst that would take less than one fifth of a second. This process makes sense, however, only when the line has some other work to occupy its time when it is not transmitting from this particular terminal. If only one terminal is attached to the line, then there is no point in having a terminal buffer (at least for considerations of timing). The need for the buffer arises when more than one device is attached to the line. In practice we can have many devices attached to one communication line, and sometimes many terminals share the same buffer. The buffers are used both for transmissions to the computer and from it.

Paper tape has traditionally been used as a cheap form of buffer. A message is punched into paper tape, and this is placed on a tape reader to be read at the computer's convenience. This scheme is used on many message-switching systems. A buffer allows messages to be composed fully before being transmitted to the computer. If the message is long or complicated, or if it takes a long time to compose, it may be worthwhile checking that it is correct before any of it is transmitted. If many long messages are sent to the computer without buffering at the terminal and the operator enters characters slowly—perhaps because she is talking to a client on the telephone at the same time—then a large amount of computer memory will be tied up in buffering the messages at the computer end of the communication lines. Partially completed messages will remain in computer storage for a relatively long period before being processed. When buffering is used, no single terminal will occupy the line for a long period of time; nor will the terminal tie up the computer for a long period.

A hardware buffer may be in the terminal itself, or it may be in a unit that controls several terminals. The logic associated with it usually carries out functions other than simple buffering.

The cost of buffering in a terminal used to be high but has fallen substantially in recent years. Nevertheless, at the time of writing, far more unbuffered than buffered terminals are installed.

ERROR CONTROL

A different reason for having a buffer in a terminal concerns error control.

As was discussed in Chapter 8, errors in the transmission may be detected by using codes designed to reveal whether any bits have been changed. Having detected an error, the receiving device may request that the item be retransmitted. The device that originally sent the message receives an instruction to resend it. This can be done only if the sending device still has it. For a terminal, a buffer is needed. Each message sent is retained in the buffer until it is known whether it had been received correctly or not.

If a terminal has no buffer, error checking of messages from the terminal may still take place, but when an error is found there can be no *automatic* retransmission. Instead the operator is notified of the error and she must initiate retransmission. The terminal must be able to send and receive the signals saying whether the data were received correctly or not.

SYNCHRONOUS TERMINALS

There are two conflicting desires: to make the terminal inexpensive and to use the communication lines efficiently by putting many terminals on one line. Until the present time, low-cost terminals have been start-stop, with no buffers. Buffered, *synchronous* terminals have been more expensive but have given better line utilization. Where the lines are short and inexpensive (e.g., within one city), efficient line utilization is of little importance. When a dial-up line is used, normally there will only be one terminal on the line and start-stop operation will often be good enough. A somewhat higher character transmission rate could be obtained with synchronous transmission (described in Chapter 7).

The other main advantage of synchronous transmission is that the error rate can be less. Extremely good error control can be achieved with high-order error-detecting codes and a buffer in the sending machine so that retransmission can be automatically requested.

The disadvantage of synchronous operation—the fact that it is more expensive—is diminishing as the cost of logic circuitry drops. At the same time, the reliability of logic circuitry is substantially increasing. All the logic for a synchronous, buffered, error-checking terminal can now be constructed on one large-scale integration chip, which can be low in cost if large quantities are mass-produced.

POLLING

A buffer in the terminal assumes a different kind of importance when more than one terminal is connected to the line, as in Figure 11.4. Suppose that terminal 4 in that diagram is waiting to transmit. However, it has to wait its turn until terminals 1 and 2 have transmitted. If terminals 1 and 2 have their messages waiting, in buffers, they will not hold up the proceedings too long. They will each transmit at the full speed of the line and with no delay. On the other hand, if the terminals do not have buffers, terminal 4 is at the mercy of the operators of terminals 1 and 2, and these operators, not even knowing that terminal 4 is waiting, will probably key in their messages with one finger.

The use of buffers, then, is the only way to guarantee a good response time on a multidrop line. Furthermore, in order to make transmission from the buffer as fast as possible, it would be worthwhile to use synchronous transmission.

A terminal on a polled line must be able to recognize its own address and must be able to respond to the polling signals. The logic for these steps will be in the terminal control unit, which may also contain the buffer, plus the ability to code and decode error-detection characters in messages and to retransmit or request retransmission when errors are found. This much is a basic logical capability found in buffered terminals.

FULL DUPLEX VERSUS HALF DUPLEX

The terminal may, as was noted in Chapter 7, use full-duplex or half-duplex transmission. It may be designed so that data can travel in both directions at once; although it is more common for data to travel in one direction while only control signals travel simultaneously in the other. Many terminals are not capable of transmitting and receiving at the same time—in which case, there is no point in paying extra for a full-duplex line for them. In some countries a full-duplex line costs the same as a half-duplex line, and so a terminal that has only half-duplex capability wastes some of the available bandwidth.

It is sometimes difficult with a terminal to organize the operation so that data flows in both directions at once—although I can imagine using a typewriter-speed terminal with a dialogue in which I sometimes respond to the computer while the terminal is still spattering out its last message to me.

With a *concentrator*, it is often worthwhile to have data flowing in both directions. Some terminals will be receiving while different ones attached to the concentrator are transmitting.

INTELLIGENCE IN THE TERMINAL

The inexpensive start-stop terminal may contain no logic other than that for transmitting and receiving each character. At the other end of the scale, a terminal may contain a high level of "intelligence." The systems analyst must assess the value of placing intelligence in the terminal *rather than elsewhere.*

If there are many terminals and one computer on a system, it makes sense

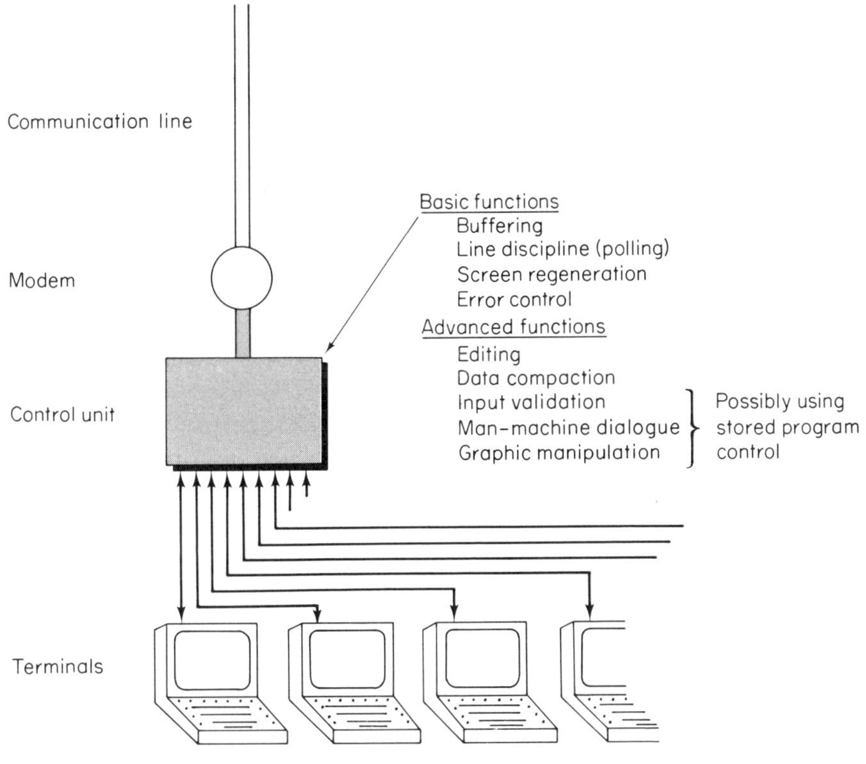

Figure 13.2

to keep the terminals cheap and to put the logic in the computer. However, where the communication lines are expensive, a new factor enters the argument: can we lower the overall communication bill by building more logic functions into the terminals?

There are other arguments beyond the question of cost. Can we create a better man-machine interface if we make the terminal more elaborate? Can we lessen the number of errors? Can we do something to lessen the inconvenience when the computer or communication line fails?

Logic capability in the terminal used to be expensive. Furthermore, it used to be somewhat unreliable, and the cost of maintenance engineers chasing round to fix the large quantity of terminals is much higher than when they are merely servicing the computers. For such reasons many unbuffered, asynchronous, simple-minded terminals are still being sold. The cost of logic, however, is dropping fast and its reliability, with the advent of large-scale integration circuitry, is becoming very high. Buffering and logic can reside on a single mass-produced silicon chip in the terminal. Like other logic devices, mini-computers are falling in cost, faster than the drop in cost of communication lines, and so the whole question of how much intelligence should reside in the terminal is being reevaluated. The systems analyst will be confronted with a much wider diversity of choice in the future.

At the upper end of the scale, the terminal becomes a stored-program computer capable of carrying on a dialogue with its user—for example, for collecting data that it then transmits to a central computer or for carrying on the manipulation of graphic images and responding to a light pen.

CLUSTERED TERMINALS

In some cases, the terminals are installed in clusters and the logic and storage discussed in this chapter may reside in a common control unit, as shown in Figure 13.2. The functions of such a control unit range from simple control of alphanumeric displays to functions requiring the capability of stored-program control.

The "intelligence" shown in the control unit in this figure may be moved farther away from the terminals and may reside in a remote concentrator. The concentrator may also be a stored-program machine.

SUMMARY

A more comprehensive summary of terminal features is given in the appendix, page 187.

14 WHAT SHOULD THE PROGRAMS BE REQUIRED TO DO?

The "intelligence" in a teleprocessing system can reside almost entirely in the central computer, or it can be scattered throughout the network. Logic functions may have their home:

1. In the terminal
2. In the terminal unit handling several terminals (Figure 13.1)
3. In the concentrator (Figure 11.5)
4. In the transmission control device attached to a computer channel (Figure 12.1)
5. In the central computer itself

Any of these logic operations performed on the data transmitted may be carried out by programming if the device in question is a stored-program machine, as all the above categories of devices can be. On the other hand, the operations may be performed by logic circuitry permanently wired into the machines or by microprogramming in which the built-in logic is modifiable.

Because of this diversity of options, the programming required for control of the transmission of data varies substantially from one selection of hardware to another. The software packages available for assisting the programmer also vary widely. On some systems he has to program for the movement of every bit or character on the communication lines: on others, software packages take care of all the functions associated with data transmission.

This chapter reviews the functions that must be accomplished to control the transmission of data, and the following chapter discusses the packaging and structure of software.

BASIC FUNCTIONS

First we will examine the basic functions that must be performed on a simple on-line system. To these a variety of other functions will then be added, which relate to more complex network control, queue control, man-computer dialogue, and so on.

Consider the basic system shown in Figure 14.1, with terminals linked to

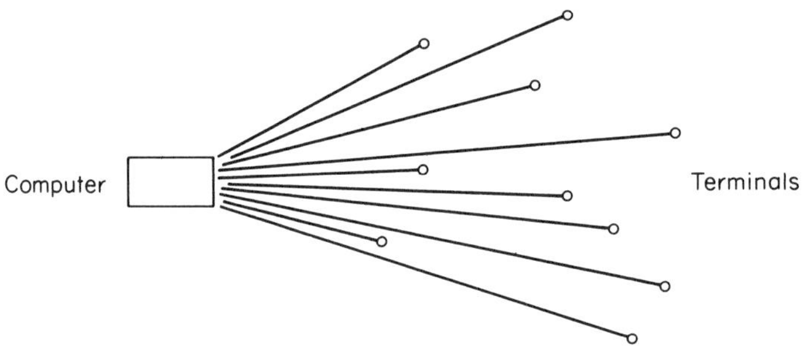

Figure 14.1

a computer by permanent point-to-point lines. With any input-output unit, such as a tape or disk, the main programs in the computer receive, or prepare for output, a complete record or message. A software routine writes or reads the record, blocking records together or unblocking them if necessary and checks that the operation has been done correctly. This procedure is also required with the communication lines, but here the input-output routine may be more complicated.

The digital information is transmitted on the communication lines a "bit" at a time. The messages to be transmitted must therefore be broken into "bits," which are sent at the speed of the line. Similarly, on receiving a message, "bits" are assembled one at a time into characters, and the characters are assembled into messages. Both the characters and messages must be error checked and the errors corrected if possible. Suitable control signals must be generated for operating the distant terminals at the correct times.

The following functions must be carried out:

1. *To initiate and control the reception* of data from the lines. The lines may be of different speeds. Many lines may be transmitting or

receiving at once. The terminals may have to be polled to see when they are ready to transmit.

2. *To assemble the bits* into characters and characters into messages.
3. *To convert the coding* of the characters. The coding for the communication lines may be different from that used by the computer. The lines may, for example, use Baudot telegraph code, whereas the computer uses binary-coded decimal.
4. *To check for errors*, both in the characters by means of a parity check (if used) and in the messages by means of block checks.
5. *To edit the messages* if necessary. For example, an operator may make mistakes in typing on the keyboard and use backspace, or erase, characters. The message must be edited into its correct format before it is ready for processing. Control characters must be recognized and removed.
6. *To recognize end-of-record or end-of-transmission characters,* and carry out the housekeeping, preparing for another transmission if necessary. If an error was detected, the same message should be sent again.
7. *To deliver messages to the main programs*, one at a time, edited and converted.
8. *To accept messages from the main programs* when they are ready for transmission to the terminals.
9. *To prepare these messages for output.* It will be necessary to convert them from computer code to communication-line code. Control characters may have to be added.
10. *To initiate the transmission* of these messages.
11. *To monitor the sending process*, repeating characters or messages if the terminal detects an error in transmission.
12. *To signal end-of-transmission* to the terminal and carry out the necessary housekeeping functions and line-control functions.

These basic functions may be done in the main computer or in an external transmission control machine. They may be done by hardware, software, or microprogramming.

POLLING AND DIALING FUNCTIONS

Software requirements increase in complexity when a more elaborate configuration of communication lines is used than that shown in Figure 14.1. The system may have polled lines, looped lines, or dial lines. The functions required to control those include the following:

13. *Dialing terminals.* The program may dial the requisite digits and establish contact with the terminal or with another computer. If the dialed station is "busy," the request for the operation may be stored and acted on later.
14. *Scanning dial-up terminals.* In some cases, a group of terminals on dial-up lines will be scanned by the program to see whether any of them have stored data to send.
15. *Answering when the computer is dialed.* When a dialed call is received from a terminal or other computer, the software should establish a connection with the device, using the requisite protocol, and should notify the necessary application program.
16. *Polling by programming.* When a line is polled, this may be done entirely by programming. The program has a list giving the terminal addresses and the sequence in which they are polled.
17. *Polling with an autopoll device.* An external polling mechanism may be used such that the software does not send the polling messages itself but rather gives instructions to the polling device.
18. *Controlling looped lines.* Where several devices are attached to a looped line as in Figure 7.8, using a synchronous stream of words, the software may assemble and disassemble the necessary bits and characters.

Once again, all these functions may be handled in the central computer or in a separate machine. For flexibility, this machine may be a stored-program computer capable of handling messages of any length, of controlling different numbers of lines, and of modifying its actions as changes are made to the network.

BASIC SOFTWARE PACKAGES

The functions just discussed constitute the basic essentials required for controlling the communication lines. Some software packages for line control contain these basic functions and little else. They relieve the programmers of the need to assemble bits into characters and characters into messages, to convert codes, and perform the other functions listed above. As we will illustrate in the next chapter, the application programmer in some cases simply writes GET and PUT macroinstructions, and appropriate coding is made available to him.

Little else is required on simple systems. On larger or more complex systems, more concern is needed with the efficiency of organization both of the central computer and the communications network. This leads to queuing schemes for messages, and higher levels of multiprogramming, to dynamic buffer allocation, to message compaction techniques, and to the communication-line concentrators, and other schemes discussed in Chapter 11.

SCHEDULING AND RESOURCE ALLOCATION

Concern for efficient use of the computers is likely to lead to the following functions:

19. *Dynamic buffer allocation.* When variable-length messages are received and transmitted, the buffers that hold them can be organized in several different ways [1]. The dynamic allocation of buffer storage in blocks can lessen the total storage requirement. The blocks will be chained together as required, and as soon as a block becomes free, it will be allocated again to the pool of available blocks.
20. *Handling line queues.* On a polled communication line, a queue of transactions can build up, if permitted, waiting for transmission. On some simple systems, no queuing is permitted; that is, no program can send a message on a line when one is already waiting to be sent. This will sometimes hold up the processing, and so a queuing mechanism is desirable.
21. *Handling queues for programs.* Similarly, queues of input messages build up, waiting for the attention of certain programs. A similar queuing mechanism may handle input *and* output.
22. *Handling items for multiple destinations.* A message may be routed not to one but to many destinations. Rather than simultaneously occupying many queues, which is wasteful, there may be one copy of the message. Control entries referencing the message would then be put into the various queue control areas.
23. *Priority scheduling.* Some messages may be given priority over others. In this case, a queue mechanism that handles multiple priorities is needed.
24. *Multiprogramming.* When applications programs have to wait for telecommunication operations to be completed, it is desirable that the central processing unit should switch its attention to other work. The ability to switch easily to lower-priority tasks and back is needed.

LOGGING AND STATISTICS GATHERING

25. *Message logging*. On some systems all messages are logged on tape or disk, for possible reference later.
26. *Statistics gathering*. An important but often neglected function is the gathering of statistics on traffic volumes, line errors, and line failures.

OTHER FUNCTIONS

A variety of other less-basic functions are also performed by teleprocessing software. Most of the remaining functions are of a more specialized nature and apply only to certain systems.

First, these are functions needed when devices are used in the network, such as remote concentrators. The programs may use a variety of techniques for blocking, compressing, or multiplexing data to be sent.

Second, there are techniques that can assist in the recovery from failures.

Third, there are features that assist in maintenance, such as diagnostic programs which an engineer at a remote terminal can use. There are procedures for checking the network and for finding out where faults lie.

Fourth, there are features related to security [2].

Finally, there are a variety of functions that the software can perform which relate to man–machine dialogue [3].

15 SOFTWARE STRUCTURES

THREE TYPES OF PROGRAM

In programming communication-based systems, three kinds of programs are needed.

1. *Application Programs*

These are the programs that carry out the processing of transactions or messages. They correspond to the data-processing programs of conventional applications and can be unique to each system. They contain no input-output coding except in the form of macroinstructions which transfer control to an input-output control routine or supervisory program.

2. *Supervisory Programs*

These programs coordinate and schedule the work of the application programs and carry out service functions for them. The supervisory programs handle input and output operations and the queuing of messages and data. They are designed to coordinate and optimize the machine functions under varying loads. They process interrupts and deal with error or emergency conditions.

3. *Support Programs*

The ultimate working system consists of application programs and supervisory programs. However, a third set of programs are needed to install the system and to keep it running smoothly. These are referred to as support programs and include testing aids, data generator programs, terminal simulators, diagnostics, and so on.

The application programs are like the workers and plant in a factory, while the supervisory programs are like the office staff, management, and foremen. The support programs are like the maintenance crew, helping to install new plant and to keep the machinery working.

Different terms are used by different organizations to describe these programs. The application programs are called "operational programs," "processors," "ordinary processors," and so on. The supervisory programs are called "control programs," "executive," "monitors," and other names. A variety of terms are used for the support programs.

OPERATING SYSTEM

In the terminology of IBM and some other manufacturers, the term "Operating System" is used for a package that includes the supervisory programs and most of the support programs. Language compilers and assemblers reside on a "system residence" disk so that they can be called in when programs arrive that need to be compiled or assembled. Application programs are stored similarly so that they too lie in wait for jobs that may need them. The operating system manages the stream of incoming jobs, whether they are from card readers or other units in the computer room or from remote terminals. The jobs are placed in queues waiting to be executed.

MULTIPROGRAMMING

A computer handling input from terminals does not follow a repetitive cycle of events as do some simple batch-processing systems. Messages arrive at random times and are often varied in their length and nature. The sequence of operations is unpredictable. However, the data are still to be handled in minimum time, and computer facilities must be used as efficiently as possible. Many systems have tight time constraints on how long the computer shall take to respond.

Furthermore, the computer will normally be handling many terminals simultaneously. Some large commercial systems have more than a thousand terminals. They will not all transmit at once, but even so the computer is bombarded with transactions from all directions. In such a case, a high degree of multiprogramming is necessitated. The computer must be like a juggler keeping many balls in the air.

A batch-processing system without data transmission may also use multiprogramming, keeping several jobs in operation simultaneously. While an input-output operation is being executed for one job, processing will be under way on another. With a batch-processing system, multiprogramming can improve the throughput. With most teleprocessing systems, however, it is a *necessity* because of the random input and fast response requirements.

TASK SCHEDULING

The nucleus of the programming system will be the top executive, which constantly determines which task shall be done next. This central scheduling routine will be returned to time and time again to make the decision: what next? The scheduling routine will have lists of tasks waiting to be done. It will select one from the list and give control to it.

Figure 15.1 shows the task scheduler. It will handle a queue of new items on which work has not yet begun, plus a queue of items in progress. The processing of the latter has been interrupted for some reason, often because the completion of a file operation was needed before processing could proceed further. Sometimes processing is interrupted by a timing mechanism—no transaction being allowed more than a certain amount of processing time at one go. The application programs will contain macroinstructions which state that their work is completed for the time being and which pass control to the task scheduler. These macroinstructions are of two types: WAIT, which says that the program must wait until certain input-output requests, usually file operations, have been satisfied, and EXIT, which says that the program is completed.

Which operation the task scheduler selects to be performed next depends on the scheduling rules that are used. There may be a priority scheme, some of the items in Figure 15.1 having higher priority than others. If not, they are handled first-come–first-served, but the new input queue might be served before the queue of partially processed items.

INPUT-OUTPUT SCHEDULING

Whereas a real-time terminal system cannot exist without a control program for handling the randomly occurring events, a batch system can—and often did in the first and second generations of computers. The reason why most third and fourth-generation computers have their "operating system" for batch work is to increase the efficiency with which the hardware is utilized.

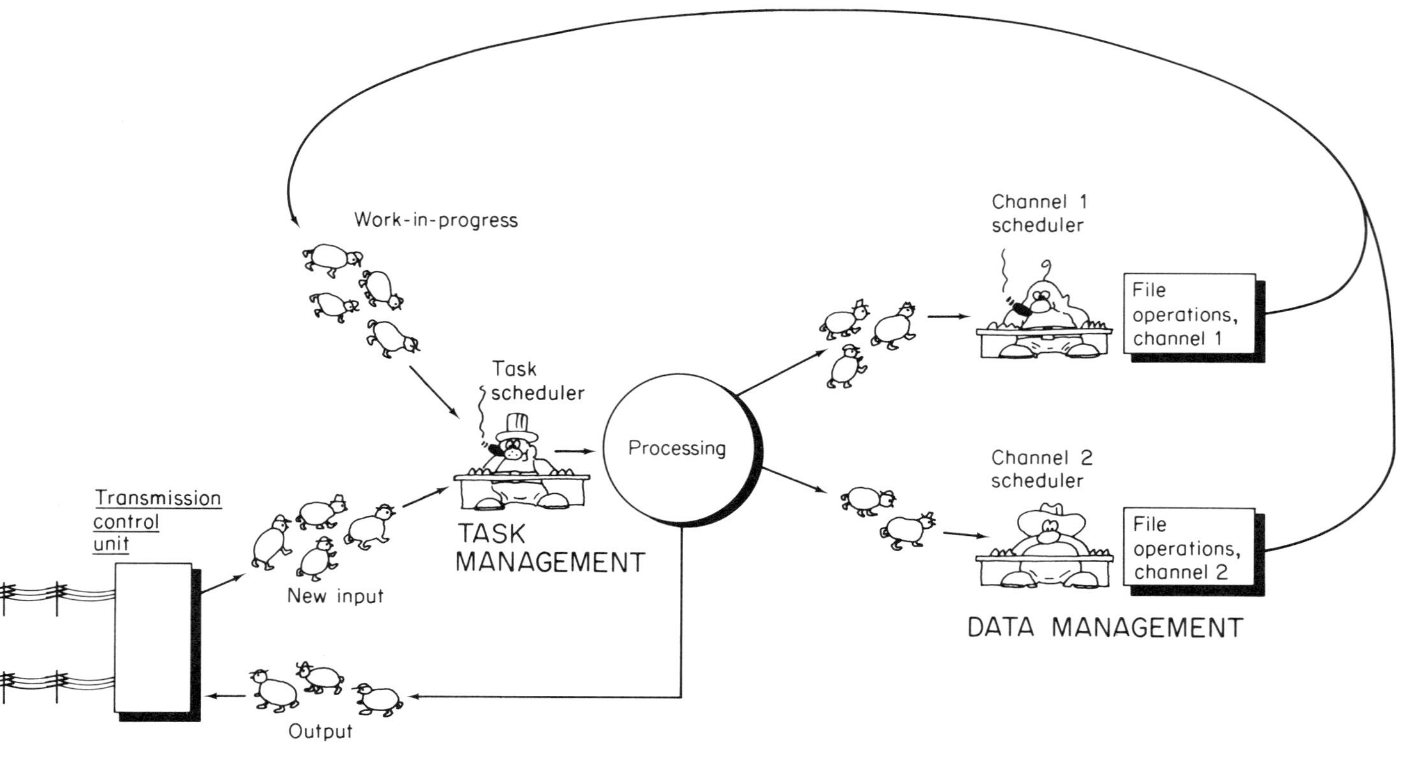

Figure 15.1. The top executive of the system is the task scheduler, which is constantly deciding which task will be processed next. Many tasks can be active at once in different phases of operation.

The task scheduler in Figure 15.1 will keep the processing unit busy, especially if there is some other work (not shown in the diagram) to use up those moments when nothing is arriving on the telecommunication lines.

Similarly, the channel schedulers, shown in Figure 15.1, keep the channels or input-output units as busy as possible. As soon as one channel operation is finished, the next one from the queue is started. As soon as one access is completed on a file unit, the next can begin. The transaction for which the file access has been completed may then be ready for processing to continue. The processing of the item had been held up, waiting that particular record from the files. The transaction now goes back into the queue labeled "work in progress," and the task scheduler will eventually give it control of the processing unit again.

When the processing of a transaction from a terminal has been completed and a response has been composed, the response is sent to that part of the software which handles the communication lines. This was described in the previous chapter. It may be in the same computer as the processing, or it may be in a separate transmission control computer. The input-output to the transmission control computer may be handled with channel scheduler routines like those shown handling the file channels.

DATA MANAGEMENT

The operating systems for the IBM 360s and 370s categorize their message handling into three functions: job management, task management, and data management.

Job management is concerned with the handling of the input job stream and is primarily designed for jobs being submitted from card decks or tapes, or from remote job entry terminals. It provides communication between the user and the operating system by analyzing the input stream and preparing a job for execution.

Task management supervises each unit of work to be done. A job is composed of many units of work, such as segments of processing, file accesses, and other input-output operations. These are called "tasks." A task scheduler, shown in Figure 15.1, maintains lists of tasks to be done and controls the sequence in which they are done. Task management is also concerned with the supervision of how main storage is used, as well as with the handling of interrupts. The main purpose of interrupts is to indicate when an input-output unit needs attention. For example, when a file access is completed, the programs running at that time would be interrupted, and a task management routine would place the item waiting for that record back in the queue of work waiting for processing (Figure 15.1).

Data management controls all operations associated with input-output devices. This includes the accessing of files, the cataloging of data sets, the allocation of space on volumes, the blocking or unblocking of records, and the handling of queues for channels and devices as in Figure 15.1. A special subset of data management handles the input-output operations associated with data transmission, which were discussed in the previous chapter.

ACCESS METHODS

The operating system *data management* software includes many different access methods. These are designed to control the input-output operations on storage devices and on telecommunications links. In so doing they relieve the programmer of much work. He no longer needs to concern himself with how the files are organized, how tape records are blocked, or how the bits are transmitted over telecommunication lines.

There are two categories of access method: those in which the input-output requests are queued, as in Figure 15.1, and those in which they are not. If they are not queued, then a program cannot begin an input-output operation on a particular facility until the previous one has been completed. If it tries, processing will be held up, waiting for the completion of the earlier operation. The ability to queue requests therefore helps to avoid keeping the processor waiting. It can give its input-output command to the data management routines and proceed with something else. Queuing, on the other hand, needs more storage and more complex programming. For this reason, it may not be used on a machine with a small main memory. On a small system there is often more concern with cost than with keeping the processor busy. With an expensive processor, however, it is important to keep it busy, and queuing the input-output operation helps to achieve this condition.

Figure 15.2 shows the access methods and the overall structure of the IBM 360 and 370 operating systems.

The application programmer uses the access methods by writing input-output macroinstructions in his programs. In the nonqueuing access methods, he writes READ and WRITE macroinstructions. In the queuing access methods, he writes GET and PUT macroinstructions. Appropriate sequences of input-output instructions are generated for these, which are executed by supervisor.

In Figure 15.2, two telecommunications access methods are shown: BTAM and TCAM. The user of the operating systems may employ one of these access methods for handling his data transmission, or he may write his own. Other organizations market teleprocessing software packages that have basic functions similar to BTAM and more comprehensive functions similar to TCAM.

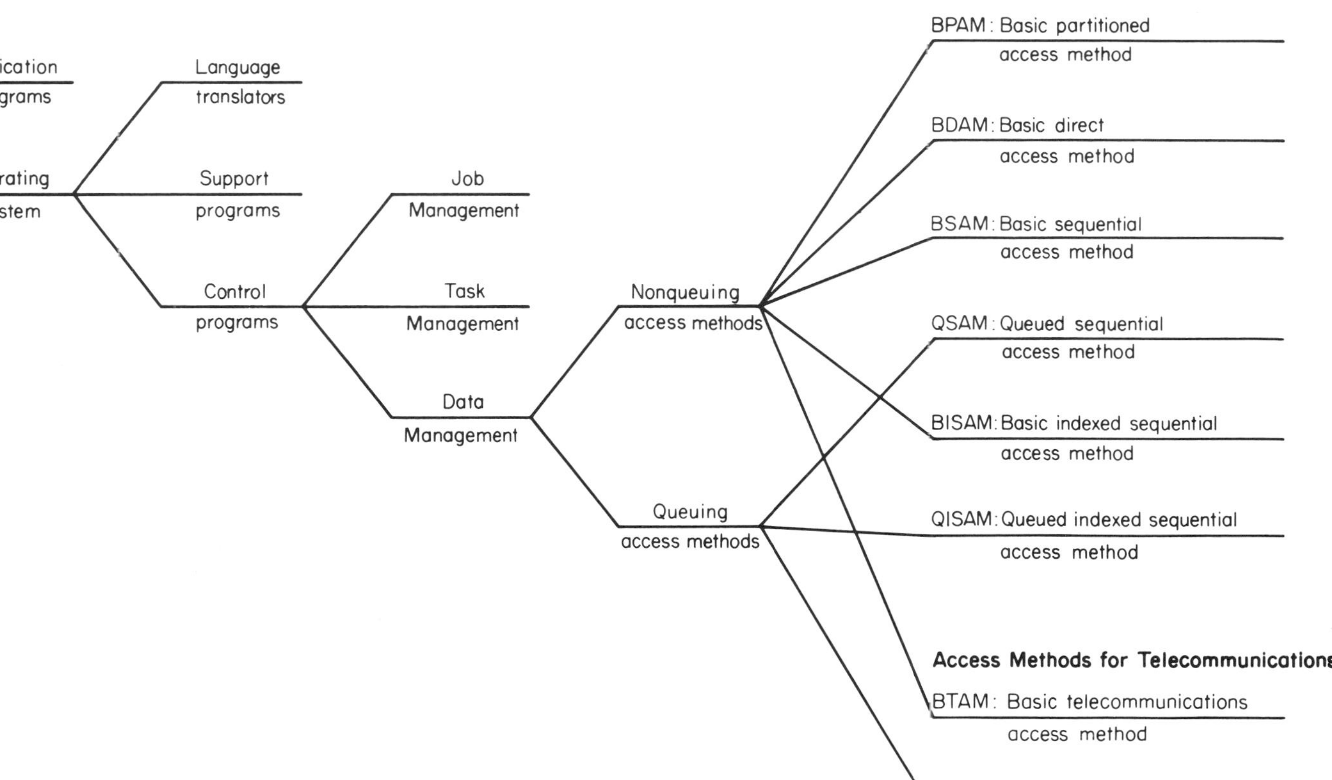

Figure 15.2. Structure of the IBM 360/370 operating system.

BTAM

BTAM (Basic Telecommunications Access Method) is a simple package typical of basic data-transmission software on other systems. It provides the basic functions needed for controlling telecommunication lines. In order to send a message to a terminal, the application programmer puts a WRITE macroinstruction in his program. To obtain a message from a terminal, he uses a READ macroinstruction. This program does not control queues of transactions, which is a disadvantage on a system with multidrop lines or a concentrator. BTAM with associated transmission control hardware carries out the first 18 functions listed in the previous chapter.

Figure 15.3 shows the linkages between BTAM and the programs using

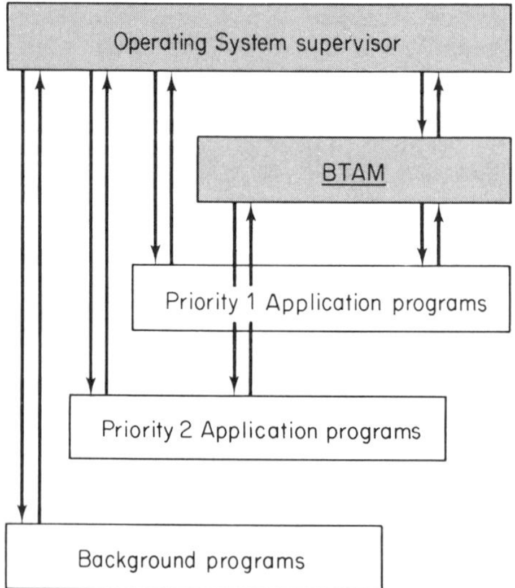

Figure 15.3. The relationship between a basic teleprocessing support package (BTAM, for example) and the application programs.

it. Two levels of priority of teleprocessing program are shown, along with a background of low-priority, nonteleprocessing programs.

This level of software support makes teleprocessing programming easier, but on complex systems the programmer is still left with difficult problems that could have been taken over by software. Some organizations have recognized this fact and have written their own software package, which fits between the application programs and the basic teleprocessing package, as shown in Figure 15.4. This particular package may reside in the highest-priority partition of the operating systems.

The reasons for writing a subsidiary teleprocessing package, like this, can be to handle some standard aspect of man-machine dialogue, to give a

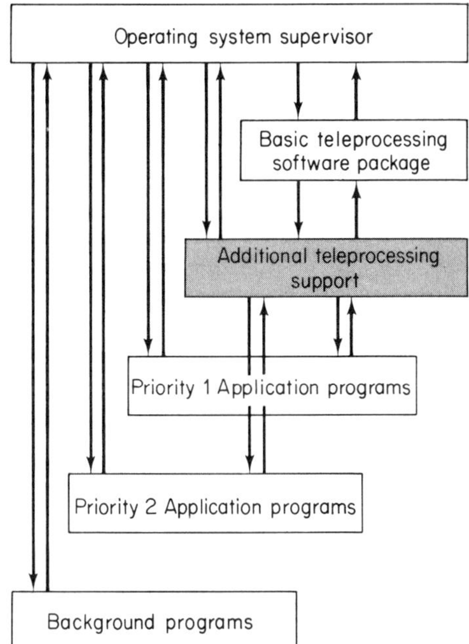

Figure 15.4. Some organizations have produced a software package that fits between the teleprocessing application programs and the basic teleprocessing software. Such a package may be written for a particular class of applications.

particular treatment of errors or failures, to further simplify the input-output coding, or (a particularly important reason) to program facilities for queuing and multiple "WAIT" conditions in the application programs so that many random events can be processed in parallel with efficient use of the processing unit.

TCAM

TCAM (Telecommunications Access Method) solves the latter problem. It maintains queues of transactions, and it can support a high degree of multiprogramming. Many transactions of different types and different categories of communications equipment may be in process simultaneously. The application programmer is insulated from the input-output operations and simply specifies them in his program with macroinstructions such as GET and PUT. He need pay no attention to the complex queuing and timing considerations.

TCAM, unlike more basic teleprocessing software, has it own *control program*, which takes charge and schedules the traffic-handling operations. Interrupts, and macroinstructions in the application programs, cause control to be given to the TCAM control program. This control program resides in one partition of the operating system and is executed as the highest-priority job in the system. The application programs are normally executed in the lower-priority partitions, although, if necessary, some may share the highest-priority partition with the control program.

Non-real-time jobs may also be run in the same system and would normally occupy the lowest-priority partitions.

Messages reaching the TCAM control program are routed by it to the requisite destination, which may be a terminal or an application program. Often the required communication line or program will be occupied, and so the TCAM control program organizes queues of items waiting for these facilities.

MESSAGE SWITCHING AND DATA COLLECTION

Sometimes TCAM can handle an incoming message by itself without needing to pass it to an application program. Such is the case when the message is merely to be routed to another terminal or computer, as in a message-switching system. TCAM, in fact, carries out by itself all the functions that would be found in a message-switching system (see Appendix, page 228). If a message from a message-switching terminal is intended for processing by an application program, the header of the message indicates this fact.

Similarly, TCAM can carry out a data collection function without reference to application programs. Terminal operators key in data that are to be collected for subsequent batch processing, rather as they might punch data into cards. TCAM will either store the data in a queue for a particular application program or will write it in secondary storage independently of an application program. In the latter case, it will be read for batch processing at some later time by an access method unrelated to TCAM.

OTHER TYPES OF STRUCTURE

Many other types of software structure are possible. The 360/370 operating systems are voluminous and complex because they attempt to meet the needs of a wide variety of users. The all-things-to-all-men software package will always be cumbersome compared with one tailored to a particular type of system.

Many control programs have been designed *solely* for data transmission. Some have been designed for message switching. Many have been designed especially for real-time systems [1]. For a real-time system, the structure shown in Figure 15.1 is often used, but it can be coded much more succinctly than with a general-purpose operating system. It may be coded to minimize the overhead used in main memory and processor time. It may be coded to give the fastest possible response time. It may be coded to handle many hundreds of terminals efficiently.

LEVELS OF TRANSPARENCY

The complexity of data transmission software in the future, however, is likely to increase rather than decrease, because of the increasing diversity of network structures and terminals.

It will become increasingly important to protect the application programmer from the complexities of the network and the terminals. In the future, we will probably see layers of software designed to make the transmission facilities as transparent as possible, whereas, in reality, they are becoming increasingly complex.

It will be important to achieve "code transparency" in a code-sensitive network so that the programmer does not concern himself with the codes used by the network or by the terminal. It is desirable that he should be able to send any combination of bits, regardless of these. It will be important to achieve "terminal transparency" such that one terminal type can be substituted for another, without affecting the application programs, or such that incompatible terminals can communicate. It will be important to achieve "network transparency" because of the increasing complexity, variety, and incompatibility of networks. The application programmer should not need to know what network is being used.

Perhaps in the future, in addition to "job management," "task management," and "data management," we shall see "network management" and "terminal management" in the software.

16 THE FLOW OF DATA

In order to consolidate the information in the previous chapters, we will now follow the events that occur when a simple use is made of a terminal.

The girl in Figure 16.1 dials a computer. She makes an enquiry with the typewriterlike terminal shown. The computer finds the information she needs and sends a response to the terminal. The entire operation takes less than a minute.

The events that take place in accomplishing this operation are as follows:

1. The girl lifts the handset on the modem, 2 (data set) and dials the computer, using a public telephone number.
2. The call is set up like any telephone call by the public telephone exchange (central office), 4. She has, in fact, dialed the location of the multiplexer, 7.
3. There are several lines going into the multiplexer. The diagram shows nine. However, for convenience of the users it is desirable that the multiplexer should always be dialable with the same telephone number. This is accomplished by a telephone company switching mechanism called a rotary, 5. When a call to the multiplexer's number reaches the rotary, the rotary "hunts" for a free line into the multiplexer. If there is a free line, the call is established; if not, a "busy" signal is given.
4. The call reaches a modem, 6, which is similar in design to the calling modem, 2. The modem, 6, is automatically placed "off-hook" as though a telephone subscriber had picked up his handset.

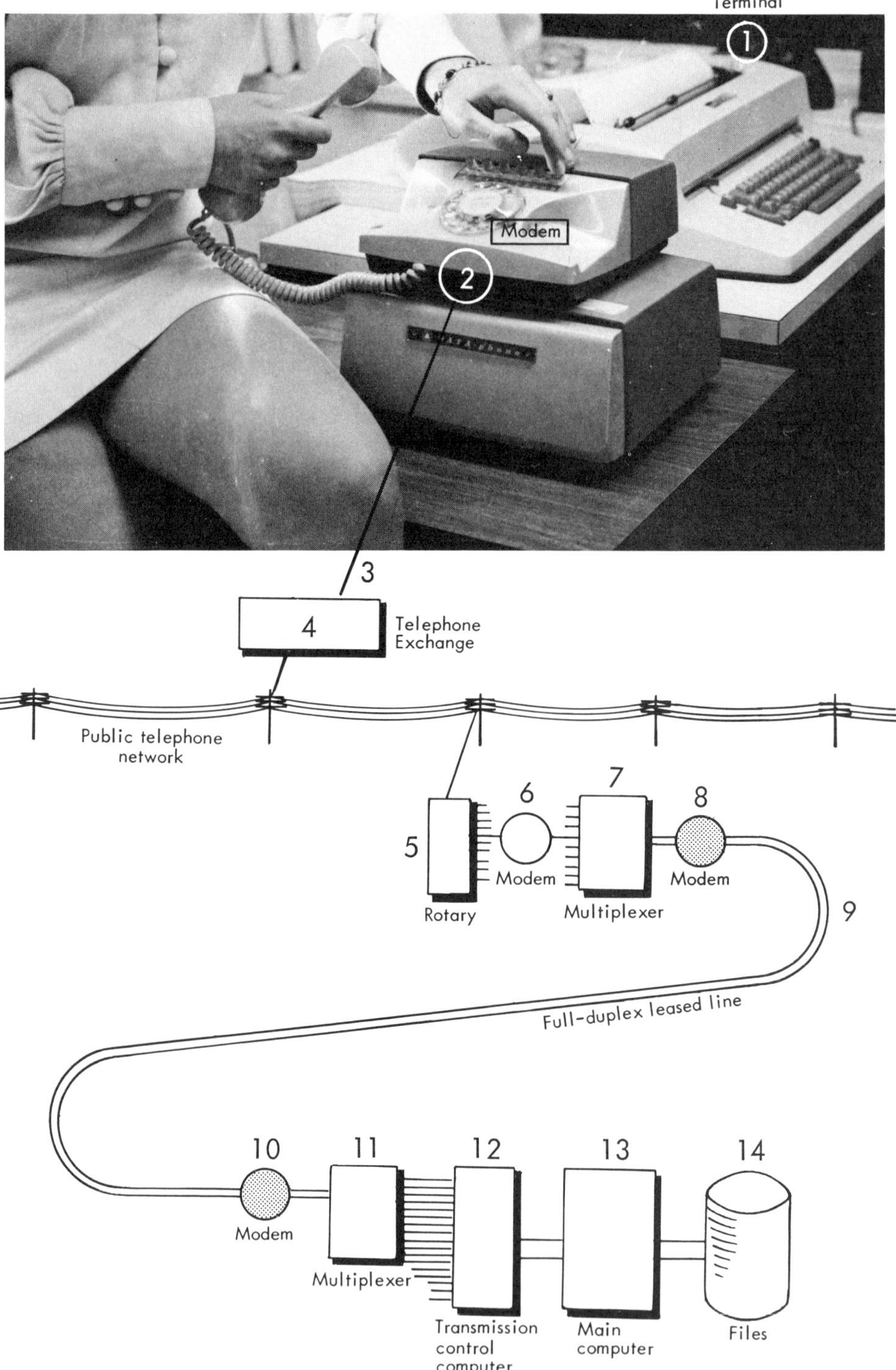

Figure 16.1. A girl at a terminal dials a distant computer to make an inquiry.

5. The modem, 6, places a high-frequency tone on the line, which means that it is ready to transmit data. The girl at the terminal hears it and presses the "DATA" key on her modem. This puts *her* modem as well in a condition in which it is ready to transmit data.
6. The girl can now type data into her terminal; the data path to the computer is set up. Modem, 2, converts the bits into signals of a suitable frequency for traveling over the telephone network, and modem, 6, converts then back into square-edged pulses, as in Figure 5.1.
7. The multiplexer, 7, constantly scans the lines coming into it, like that in Figure 11.1. It takes one character from each line and assembles them into words, which it sends onward to the computer on a leased line. The transmission from the terminal, 1, to the multiplexer is start-stop (Figure 7.4)—the terminal has no buffer or synchronization circuits. The transmission between the multiplexers, 7 and 11, is synchronous (Figure 7.6) so that the most efficient use can be made of the line linking them.
8. The multiplexers are linked by a leased, conditioned, voice-grade line, 9. Modems, 8 and 10, convert the digital signals so that a high bit rate can be achieved over this line. The modems are more sophisticated in their design and more expensive than modems 2 and 6. They transmit at 4800 bits per second, whereas modems 2 and 6 can achieve up to 300 bits per second only.
9. Multiplexor 11 demultiplexes the incoming signals in an exactly converse manner to multiplexer 7. The lines leaving multiplexer 11 and the signals on them are essentially the same as those entering multiplexer 7. The lines entering the transmission control unit 12 could, in fact, have been connected directly to the terminals. The multiplexers and other equipment between the terminals and the transmission control unit are said to be "transparent," and the programmers are not affected by their existence. With some network schemes, this is not the case [1].
10. The transmission control computer assembles the bits it receives into messages. The terminal, 1, composes sets of error-detection bits to transmit at the end of each message. The transmission control computer checks these and sends an acknowledgement to the terminal, saying that the message was received correctly. If it was not received correctly, the transmission control computer will request retransmission.

11. The computer could have been dialed from a terminal anywhere, and for security reasons the transmission control computer establishes the identity of terminal, 1. The terminal has built into it the capability to transmit an identification number on request. The transmission control computer requests this number and the terminal transmits it.
12. When the transmission control computer has a complete, checked-out message for the main computer, it interrupts the main computer. The supervisor that responds to the interrupt initiates the reading of the message into the computer's storage.
13. The message awaits the attention of the task scheduler (Figure 15.1) and is eventually transferred to an application program for processing.
14. One or more file accesses may be required, and these are handled by the data management software (Figure 15.1). Eventually a response for the terminal is composed and sent to the transmission control computer, again by data management software.
15. The transmitting of the response to the terminal is the converse of the input transmission. It goes via the multiplexers and modems 10 and 8. Line 9 is full duplex, and the transmission to the terminal takes place while other messages from other terminals may be arriving on this line.
16. Multiplexer 7 forms start-stop characters to send to the terminal. The switched line connection via the rotary and the public telephone exchange, 4, are still set up. The terminal types the response.
17. The girl reads it, and, if satisfied, she breaks the connection by pressing the VOICE key on her modem, 2, and replacing the telephone handset. The circuits are then freed to accept other calls.

Where there are several terminals on one line or where polling or concentrators are used, the control of the transmission will be more complex than what has just been described. Examples of more elaborate line control are given in the author's *Teleprocessing Network Organization*.

17 DESIGN TECHNIQUES

The designer of a data transmission system has many alternatives open to him —more, indeed, then this introductory book has indicated. This is especially so if there are many terminal locations.

Most of the design decisions are susceptible to precise calculations, although the calculations become complex in some cases.

The objective of the design is to minimize the cost of the system within certain constraints. Typical of the constraints that will be applied are (1) there should be a very low number of errors, (2) the probability of failing to obtain a connection should be suitably low, and (3) the response times should be suitably fast.

The designer, in fact, will be concerned with the *risks* that are listed on page 134. In some cases, he will say that no risk can be tolerated. He may, for example, decide that *every* call must establish contact with the computer center; there must be no "busy" signals. In this case, a leased-line network must be used and there must be no line-switching components or other devices that could block a call. Multiplexers and certain types of hold-and-forward concentrators *can* be used, for they do not block calls. This no-risk decision has been made on airline reservation systems and many other commercial systems in which sudden inaccessability of the computer would cause a problem. If the designer really means *no* risk in this case, he must go further and protect his system from leased-line failures. This means that alternate paths must be available between points in the network, along with the ability to switch between them. This is much more expensive than a simple non-blocking network.

In fact, as in other walks of life, designs for which there is *absolutely* no risk are impossible. Designs for which the risk is extremely small are extremely expensive. The designer is seeking a compromise—a reasonable degree of safety at a reasonable cost.

The assessment of the various risks cannot, therefore, be made in absolute terms. It must be made in terms of probability. And so in the design of data transmission systems, probability calculations are used, in various different forms. This has long been true in the design of telephone facilities. If all the subscribers on a telephone exchange attempted to place a call at once, most of them would not be connected. The problem becomes one of giving a certain probability of being connected under normal operating circumstances. Having established this probability, the rest of the exercise is concerned with achieving it at the lowest cost.

A similar consideration is true with the question of response time—the time taken to obtain a reply at a terminal. On many systems it is desirable to have a low response time [1]. If there is only one terminal on a leased line, an examination of the line discipline and turnaround times will show precisely how much the communication facility adds to the response time. However, if there is more than one terminal per line, then a specific determination of response time is not possible and the question becomes: what is the probability of one terminal being kept waiting by the other terminals? This probability will be greater, the greater the number of terminals attached to the line. It is possible to evaluate the *mean* response time and its *standard deviation.* Both will increase as the volume of traffic increases. To understand how the network design affects the behavioral characteristics of the terminals, the mean and standard deviation of response time should be plotted against traffic intensity [2].

On some systems the time to establish a connection is important, as well as the response time, once the connection is set up. This again must be assessed in terms of probabilities.

Often a lengthy sequence of steps is necessary in designing a data transmission network. For a large network the geographical layout of the lines can be a complex process, but before this is done, decisions have to be made about the techniques to be used on the lines—multiplexers, exchanges, concentrators, and so on. Calculations must be made about the permissible loadings of whatever facilities are contemplated.

Figure 17.1 shows a network with more than a hundred cities connected to a computer at Chicago. There are many possible ways to interconnect these cities. The one shown is certainly not the lowest in cost. Much cheaper networks could be designed if each low-speed line went to more than one termi-

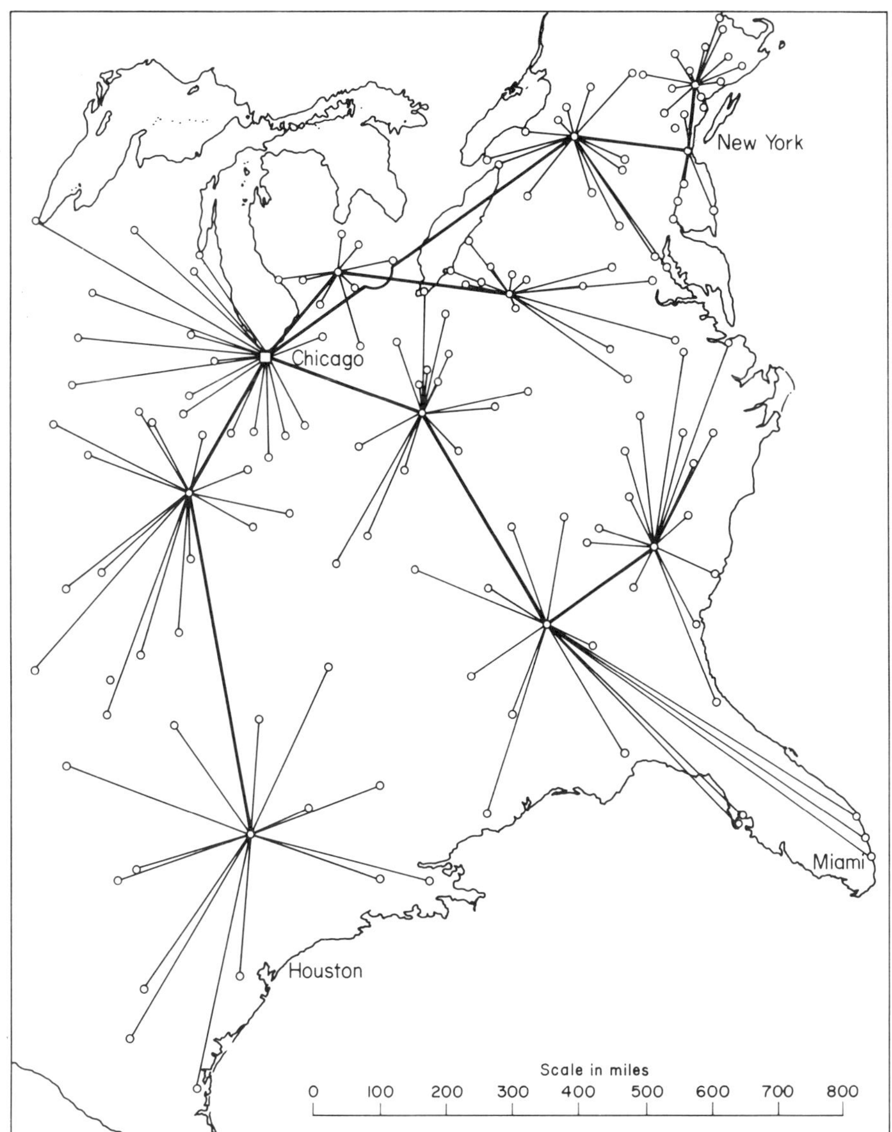

Figure 17.1. Terminals connected to Chicago by means of a network with concentrators. It is necessary to lay out a minimum-cost network that meets the design constraints.

nal. To do so, however, would result in response-time degradation. How serious this would be needs to be calculated. The network in Figure 17.1 is a minimum-cost network for one particular set of constraints.

The network in this figure poses many questions. First, are the concentrators shown at the best locations or could the network be cheaper if different locations had been selected? Could more concentrators be attached to a high-speed line? Has the best type of concentrator been selected, or could a lower overall cost be obtained with a cheaper concentrator? Or a more expensive concentrator? Could the low-speed lines each connect more than one terminal to a concentrator? If so, which terminals should the low-speed lines interconnect? There are different traffic volumes from the different cities, and an immense number of different multidrop line configurations are possible.

The design procedure will often be an iterative process. At first, a guess may be made at a possible network. Calculations will be done to see whether this fulfills the requirements. If it does not, the system will be adjusted until it does fulfill them. If it does, alternative approaches will be tried with a view to minimizing the network cost. Once a particular form of network has been decided on, various computer algorithms are available for establishing an optimum configuration of that network.

Figure 17.2 shows a possible sequence of steps in the design of a data transmission network for a real-time system. This sequence may be only one step in the larger process of designing the real-time system [3].

The process begins by determining what types of messages are to be transmitted. Then, for each type, it is necessary to determine the volumes of traffic at each location. The volumes may vary with the time of the day and the day of the year. This variation must be investigated and the peak loads established.

Step 3 in Figure 17.2 is the establishment of the response-time criteria. This can have a major effect on the types of devices that are permissible on the communication lines.

Next we must establish where the terminals are going to be. Some systems have little or no scope for discussion on this point. A terminal is located in each sales office, for example. On other systems it is a major question. The airlines have terminals in a limited number of locations, not in every city where bookings are made. If you telephone Pan American in Boston about your flight, your call is routed to girls at terminals in New York, on TELPAK lines shared with other airlines. Determining whether to put terminals in Boston, or other cities, is a major cost trade-off calculation.

Similarly, we must determine how many terminals are needed in each location to handle the traffic load. There may be queues of persons requiring service from the terminal operators. They may be physical queues as at a bank counter, or queues of persons waiting on the telephone becoming impatient at the supposedly soothing tones of a prerecorded voice. There may be no queues at all; people simply go away if some types of terminal are not available when they need them. Again probability calculations may be used.

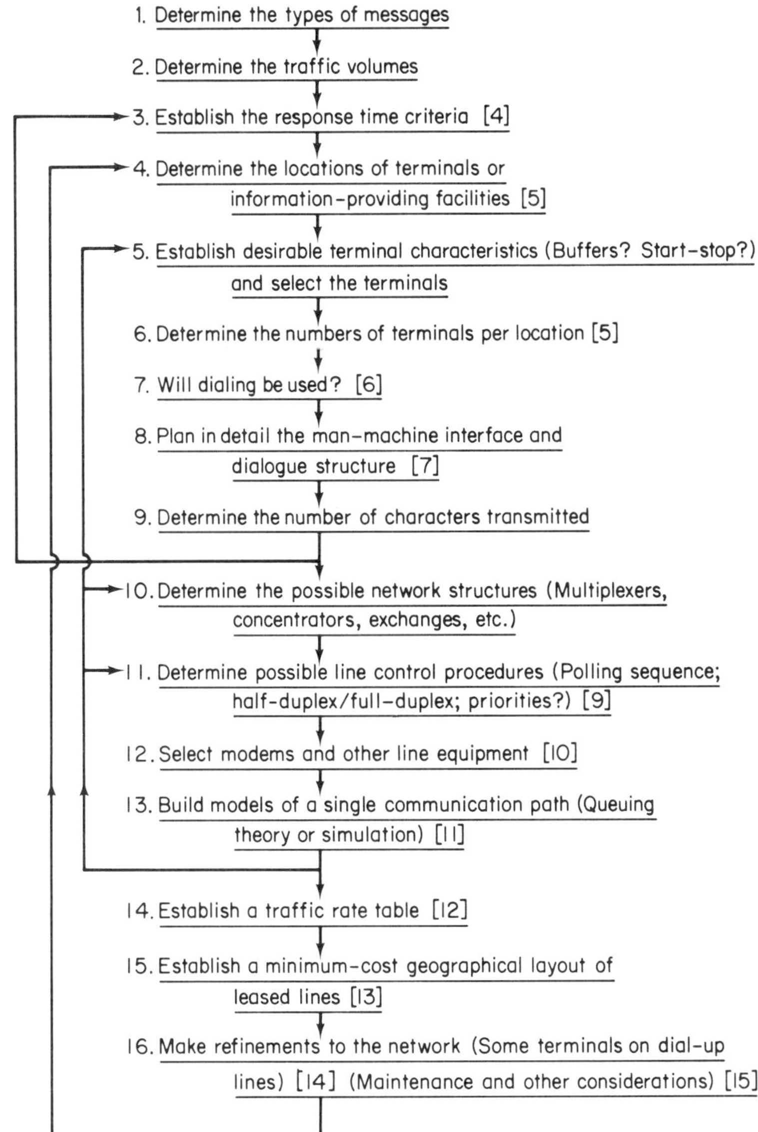

Figure 17.2. Designing a complex data-transmission network is usually an iterative process. The steps shown above are typical for a real-time system.

Step 5 of Figure 17.2 is to determine the desirable terminal characteristics. Those that affect response time are the speed of output (typewriter, higher-speed printer or screen), whether or not the device has buffers, whether it is designed for use on full-duplex lines, what control characters it needs in the transmission, whether it can be multidropped on a line, and if so how it responds to polling, and how long it takes to turn around the direction of transmission.

Step 7 asks whether dialing will be used from the terminal to the computer. If dialing is used, it is necessary to decide on a figure for the probability of failure to establish the connection. This enables the numbers of computer ports, tie lines, paths through the exchanges, and other facilities to be determined. Sometimes it is quite unacceptable to use dialing because of the delay involved. When dialing *is* acceptable, it is sometimes economical to have some of the terminals on dial-up and some on leased lines.

Before we can examine the operation of the line in detail and determine such factors as how many terminals we can place on it, the lengths of the messages must be determined. We require a distribution of message lengths. If the system is one in which a conversation takes place between computer and terminal user, the structure of the conversation must be known in some detail.

When the message lengths are known, it may be necessary to reiterate the design. The cost may be too high. It may be desirable to slacken the response-time requirement or change the terminal equipment to give more suitable facilities.

Various approximate calculations may now be done to evaluate whether devices such as multiplexers, private exchanges, or concentrators could be used on the lines. The various possible means of lowering the network cost should be checked over. At the same time, the modems may be selected. It is important to pay attention to modem turnaround time as well as transmission speed on a multidrop line.

At this stage, "models" of a single communication path, from terminal to computer, may be examined. These may be mathematical models using queuing theory or they may be simulation models. The communication line can be simulated relatively simply. The response times indicated by the models will be examined. A plot should be produced of how the mean and standard-deviation response times vary with throughput. The models may be adjusted to investigate the effect of differing the line organizations and differing the characteristics of the terminals, modems, and other equipment.

The line models that are built, whether using queuing theory or simulation, will make it possible to produce a traffic rate table (step 14 in Figure 17.2) which states the maximum traffic loads that are permissible with different numbers of terminals on the lines.

The geographical layout can then begin. With multidrop lines spanning many cities, there are an immense number of different ways of interconnecting the terminals. With concentrators, multiplexers, and private exchanges, the problem becomes much more involved. A variety of different algorithms and techniques are in use for laying out data transmission networks.

For further information on this complex subject, the reader is referred to the author's *Systems Analysis for Data Transmission* in which the steps in Figure 17.2 are expanded upon in detail.

APPENDICES

SUMMARY: THE SYSTEMS ANALYST'S DECISIONS

A systems analyst or system designer is faced with many alternatives when planning a system that employs data transmission. A failure to understand all the choices, or to know that they exist, has often resulted in a system being designed that is less efficient or more expensive than it need be. This appendix consists of summary charts that list the major design alternatives and indicate their possible advantages and disadvantages. Many of the items have been discussed in this book, but some are dealt with in greater length in my other books. The right-hand column of the charts directs the reader to the book and chapter that best discusses the choice to be made.

The books, all published by Prentice-Hall, Inc., Englewood Cliffs, N.J., are referred to with the following abbreviations:

DESREAL: *Design of Real-Time Computer Systems*
DIALOG: *Design of Man-Computer Dialogues*
FUTCOM: *Future Developments in Telecommunications*
INTROTEL: *Introduction to Teleprocessing*
NETWORK: *Teleprocessing Network Organization*
PROGREAL: *Programming Real-Time Computer Systems*
SECACC: *Security, Accuracy, and Privacy in Computer Systems*
SYSTRAN: *Systems Analysis for Data Transmission*
TELCOM: *Telecommunications and the Computer*

CONTENTS OF THIS APPENDIX

A Terminal Considerations

1. Manual-input facilities
2. Document-input facilities
3. Output facilities

4. Features of display screens
5. Features for security
6. Features for control of errors
7. Features of the communication line interface
8. General

B Man-Machine Interface Considerations

1. Categories of terminal operator
2. Response-time requirements
3. Types of approach to man-computer dialogue

C Communication Lines

1. Categories of communication line and tariff
2. Public (dial-up) line types
3. Leased-line types

D Transmission Techniques

1. Modes of transmission
2. Techniques for controlling line errors
3. Criteria for choice of modems
4. Modems: types of devices
5. Modems: theoretical performance data

E Network Structures

1. Methods of attachment of communication lines to the main computer
2. Types of leased-line network structures
3. Basic multiplexer features. A checklist.
4. Features of "intelligent" concentrators. A checklist.

F Software

1. Summary of features in data transmission software
2. Functions of message-switching software

A TERMINAL CONSIDERATIONS

1. *Manual-Input Facilities**

   Typewriterlike keyboard
   Keyboard with letters in alphabetic sequence
   Keyboard like Touchtone telephone
   Keyboard like calculating machine
   Matrix keyboard
   Keys with special labels
   Keyboard with interchangeable key labels
   Keyboards with overlays or templates
   Lever set
   Rotary switches
   Pushbuttons
   Light pen with display tube
   Coupled stylus on "desk pad"
   Pen-following mechanism
   Badge reader
   Punched-card reader
   Matrix card holder

Can the data be keyed into a buffer and modified before transmission?
Is paper tape or any other serial medium used for buffering?
Does the keyboard give any help in formatting messages?
Does it have the necessary keys for the dialogue in question?

*Book: SYSTRAN, Chapter 12.

Does it have special facilities for when the computer fails?
Can it be operated with one hand?
Can the key labels be changed?
Does a bell ring at the end of a line?
Are there good cursor controls?
Are there facilities for easy modification of computer data?
Are there skip and tab keys?
Are there page or scroll keys?
Are there YES/NO or other keys for high-speed scanning?
Is the manual correction of errors easy?
Can the numeric part of the keyboard be operated by one hand? (3 x 4 matrix)
Does the keyboard have a good "feel" to a fast touch-typist?
Are HELP or INTERRUPT keys desirable in the man-machine dialogue?
Does the input means have appropriate security features?

2. *Document-Input Facilities**

Paper-tape reader/punch
Card reader/punch
Magnetic-tape cassette
Disk

Does the machine in the preceding four cases have the facility to write or punch a document as it is being keyed in?
Can one storage media be shared by many keyboards?

Badge and credit card (identity card) reader
Optical document reader
Magnetic-ink document reader
Mark-sense card reader
Matrix plate or card reader
Magnetic card reader

Can various devices be attached to one control unit?

3. *Output Facilities**

Typewriterlike printer
Printer that operates faster than a typewriter
Inexpensive numeric-only printer, like a calculating machine

*Book: SYSTRAN, Chapter 12.

Should the printer print on a special document like a bank passbook?

Should the printer have special characters—for example, for mathematics, text editing, chemical formulas, or producing diagrams?

Should the character set be interchangeable as with an IBM Selectric "golf-ball"?

Is hard copy essential or would it be possible to do without it?

Could the hard-copy facility be at the computer or concentrator rather than at the terminal?

Could a camera be used rather than an expensive copying machine?

If a printer or plotter is needed, could one such device serve many terminals?

Visual display screen (Features of visual displays are listed separately below.)

Picturephone

Interface to standard television set

Light panel

Graph plotter

Strip recorder

Telephone voice answerback

Dials

Facsimile machine

Projector for slides, microfilm, microfiche, EVR frames, etc.

How rapidly must the data be printed or displayed? This point relates to the delivery time for bulk transmission and to man-machine interaction processes in a dialogue system.

Is there a means of alerting the operator's attention?

Can certain fields be highlighted—for example, with color?

Does it have appropriate tabbing, skipping, and page-change features?

4. *Features of Display Screens*

Can it display enough characters?

Are the displayed characters large enough?

Are the characters easy to read?

Is it flicker-free?

Is the image bright enough? Some displays have caused operator headaches.

Is the image suitably protected from external glare?

Is the display rate fast enough for the man-machine dialogue?

Can it handle vectors or other graphic features?

Destructive or nondestructive cursor?

Can the cursor be made either destructive or nondestructive under program control?

What cursor movements are possible?

Are the keys for cursor movement straightforward?

What character insert and delete capabilities are available?

Does it have suitable special characters?

Is the character set large enough?

Should the character set be interchangeable (e.g., with microprogramming)?

Does it have a scroll feature (text roll up and roll down)?

Does it have selectable horizontal tabs or other formatting features?

What editing capabilities are available?

Can individual fields be highlighted in some way—for example, by blinking, color, different brightness, or reverse field (either black characters on white or white characters on black)?

Are the features changeable—for example, microprogrammed?

Are spaces to the right of an end-of-line character transmitted?

Does it have line addressing so that part of a display can be changed without the rest?

Can a protected field be used for selective data entry?

Are upper- and lower-case characters needed (e.g., in text editing)?

Should images be displayed that are not composed digitally (e.g., documents, signatures, photographs, diagrams, etc.)? These may be stored at the terminal on film, microfilm, slides, EVR cartridges, etc.

Can such images be half-tone (like a photograph)?

Can such images be in color?

If locally stored images are displayed, should they be combined with transmitted data?

Is a group display needed?

Is an extra-high capacity screen needed (10,000 characters or equivalent)?

5. *Features for Security**

Unique terminal identification by the computer

Lockable keyboard

Nonprinting feature for keying in security code

Identification card reader (e.g. for magnetic stripe credit or ID card)

Cryptography feature

Physical lock and key

6. *Features for Control of Errors*†

Erase, or backspace, key

Cancel transaction key

Automatic error detection. The code used for automatic error detection can

*Book: SECACC.

†Book: SYSTRAN, Chapters 15 and 35.

range from relatively insecure parity checks to virtually error-proof polynomial codes of a high order.

Automatic transmission when an error is detected (which implies some form of buffer at the terminal)

Forward-error correction

Transaction logging in the terminal

Accumulators in the terminal for keeping totals

Logging facilities and/or accumulators that record details of transactions entered when the computer is inoperative

A recording mechanism (e.g., tape cassette) for recording transactions when the computer is inoperative, which can later be transmitted to it. This transmission may or may not be automatic.

7. *Features of the Communication Line Interface**

Is a standardized code used (e.g., ASCII)?

Is a compact code needed to maximize transmission efficiency?

Can *different* codes be used, for flexibility?

Can *any* characters be used (e.g., with an escape character mechanism)?

Synchronous or start-stop operation?

Is a buffer used so that transmission can be at maximum line speed (important, as we shall see on high-performance multidrop lines)?

Can operator editing of input be done before transmission (especially with a video display)?

Is the transmission rate suitably high?

Is the transmission rate changeable for bad line conditions?

Is a modem or acoustical coupler built in?

If not, does it have a standard EIA RS232B (or other) interface with the modem?

Does it have multiplexing facilities—for example, a modulation device selecting a set portion of the available bandwidth?

Can several terminals be connected to one buffer or control unit?

Does the control unit dynamically assign buffer space?

Can there be terminal-to-terminal communication?

Does it contain logic for multidrop operation (many terminals on one line) such as polling?

If polling is used, is it roll call or hub?

If one device on a multidrop line fails, will this affect the line functioning?

What communication line turnaround time is associated with the terminal?

How does the line turnaround time affect the network organization and response time?

*Book: SYSTRAN, Chapter 13.

Is full-duplex or half-duplex transmission used?

If it is full duplex, is it designed so that full advantage can be taken of simultaneous transmission in both directions?

Is there any form of interrupt mechanism?

Does the terminal have dial-up capabilities?

Can it dial a remote machine automatically?

Will it automatically redial if it first obtains a busy signal?

If it is normally attached to a leased line, can it have an alternate dial-up connection, possibly at lower speed?

Will it automatically establish a dial-up connection if a leased line fails?

Will it automatically dial a public line if a WATS line is busy?

Is the terminal monitored for running out of paper or other holdups?

Can unsolicited messages be sent to an idling terminal?

Can an idling terminal be dialed?

8. *General*

Is it portable?

Need it be battery operated (e.g., when a terminal is taken in a car and used in a public call box)?

Is it silent?

Is it compact enough?

Is it attractive?

Is it robust?

Is it reliable?

Is it designed so that it can be easily serviced?

Is the maintenance contract good enough?

Is the maintenance organization good enough?

Are there replaceable blocks of components for quick repair?

Can these blocks be changed by trained local staff so that a visit from a repairman is unnecessary?

Can the remote computer be used in terminal fault diagnosis or checkout?

Is the construction modular so that separate keyboards or other devices can be used?

Can the device be used off line for a needed function (e.g., typewriter or desk calculator)?

B MAN–MACHINE INTERFACE CONSIDERATIONS

1. Categories of Operator the Dialogue may be Designed for

			Operator with Programming Skills		Operator with High IQ		Operator without High IQ		
			With detailed training	With little training	With detailed training	With little training	With detailed training	With little training	Totally untrained operator
			1	2	3	4	5	6	7
Dedicated	Active Passive Intercept Intermediary	A B D E							
Casual	Active Passive Intercept Intermediary	F G I J							

Book: DIALOG, Chapter 3.

2. Response-Time Requirements

	Advantages and Disadvantages	Book	Chapter
Almost instantaneous	Needed in pseudomechanical coupling such as the appearance of a curve in a screen following quickly after the light-pen motion that "drew" it.	SYSTRAN	7
	Usually not achieved over transmission links but from hardware close to the terminal.		
Less than 2 seconds	Necessary in many man-computer dialogues with continuity of thinking.	SYSTRAN	7
	Needs appropriate choice of network equipment and organization. Needs appropriate control program procedures.	SYSTRAN PROGRT	
2 to 4 seconds	Longer than 2 seconds can be inhibiting to terminal operations demanding a high level of concentration. It is OK for simple inquiries.	SYSTRAN	7
	Some forms of polling and line-control procedures should be avoided.	SYSTRAN	36
	As with the above response time, careful attention to network design calculations is needed.	SYSTRAN	28
Greater than 4 seconds	Generally too large for dialogues in which the operator makes a major use of her "short-term memory."	SYSTRAN	7
Greater than 15 seconds	"The system had better be designed to free the user from physical and mental captivity, so that he can turn to other activities and get his displayed answer when it is convenient to him."	SYSTRAN	7
	Makes possible some inexpensive network structures using multidrop low-speed lines.	SYSTRAN	40

Book: DIALOG, Chapter 17.

3. Types of Approach to Man-Computer Dialogue

	Advantages and Disadvantages	Book	Chapter
1. Programming languages	Advantages: Concise, precise, powerful. Disadvantages: Inappropriate for the vast number of terminal users who are not programmers. Too difficult for the majority of people who want to carry out relatively simple operations at terminals.	DIALOG	5
2. Natural-language dialogue	Advantages: The best theoretical man–machine interface. Disadvantages: Natural-language input is of little practical value today because of ambiguity and imprecision in our language. Immense software problems. "The English we speak takes for granted an immensely elaborate web of interconnections between words that are used. Human beings grasp the meaning of sentences because they have an internal encyclopedia to which the sentences are relevant."	DIALOG	4
3. Limit English input	Advantages: User employs words he is familiar with. Disadvantages: Some users tend to overestimate the intelligence of the machine and overstep the tight restrictions on input wording.	DIALOG	7
4. Dialogue using mnemonics	Advantages: Can be concise and precise (e.g., airline reservation dialogue). Disadvantages: Operator must be familiar with mnemonics and formats.	DIALOG	7
5. Dialogue with program-like statements	Advantages: Can be concise and precise. Disadvantages: Operator must be well trained, familiar with the coding, and have a limited programming aptitude.	DIALOG	7

Dialogue approach	Advantages and Disadvantages	Book	Chapter
6. Computer-initiated dialogues (in which the operator responds to the computer rather than the computer responding to the operator)	Advantages: The computer tells the operator what to do: little training required. Can be used with a totally untrained operator. Disadvantages: Dialogue can be lengthy and often slow. Many characters, high-line utilization, more expensive networks. Little flexibility in sequence of operation.	DIALOG	6 and 7
7. Form-filling (in which the operator fills out a "form" on a visual display)	Advantages: Straightforward for the operator except for cursor manipulation. Disadvantages: Less flexible than a "branching tree" of questions, and error-correction procedures are less easy.	DIALOG	7
8. Build dialogue features into special terminal hardware	Advantages: The intent is usually to clarify and simplify the operator actions. A similar effect could often be achieved without special hardware, however. Disadvantages: Expensive. Inflexible. May restrict future development.	DIALOG	9
9. Fixed-frame responses (in which the computer responds with one of a standard set of frames)	Advantages: Simple for programming. The frames can be stored in devices away from the main computer giving low-transmission requirements. In some cases, frames with pictures may be used. Disadvantages: Inflexible. Of limited scope.	DIALOG	7, 13 and 15
10. Dialogue via a third party	Advantages: Many important uses (information room, data secretaries, telephone agents, counter clerks, etc.) Enables management, and the general public, to obtain information from computers. Disadvantages: Generally prevents extended use of the terminal.	SYSTRAN	11

There are many variations on the different dialogue approaches. For detailed pros and cons, see *Design of Man-Computer Dialogue*.

C CATEGORIES OF COMMUNICATION LINE AND TARIFF

Types of Link	Comments	Book	Chapter
Digital link	Designed for digital transmission. No modem required. Are code-sensitive in some cases.	FUTCOM	6
Analog link	Transmits a continuous range of frequencies like a voice line. Modem required.	TELCOM	10
Switched public	Cheaper if usage is low. Switched telephone lines are universally available.	SYSTRAN	18 & 41
Leased (sometimes called "private")	Cheaper than public lines if usage is high. May have lower error rate. Higher speeds possible on leased telephone lines than switched ones.	SYSTRAN	18 & 41
Leased with private switching	May give the lowest cost. Combines the advantages of leased lines with the flexibility of switching. Public switched wideband lines may not be available.	NETWORK SYSTRAN	8 33
Private (noncommon-carrier)	Usually only permitted within a subscriber's premises. See next item.		
Private (noncommon-carrier) links:			
In-plant	Very high bit-rates achievable.	NETWORK	
Microwave radio	Permissible in special cases for point-to-point links.	TELCOM	8
Shortwave or VHF radio	Used for transmission to and from moving vehicles or people.	FUTCOM	13
Optical or infrared	Used for short links—e.g., intercity—at high bit rates (250,000 bps, typical). No license required. Put out of action by fog or *very* intense rain.	FUTCOM	13
Speeds			
Subvoice grade	Usually refers to speeds below 600 bits per second.	TELCOM	5 & 8
Voice grade	Usually refers to analog voice lines using modems of speeds from 600 to 10,500 bits per second.		
Wideband	Speeds above those of voice lines, most commonly 19,200; 40,800; 50,000, and 240,000.		
For a detailed list, see the following table.			
Mode of operation			
Simplex	Not normally used in data transmission.		
Half duplex	The most common in the United States.		
Full duplex	On the Bell System costs 10% more than half duplex and can give disproportionately higher throughput if the terminal is designed to take advantage of it.		

NOTE: These terms sometimes describe the limitation of a machine rather than the limitation of the line it is attached to.

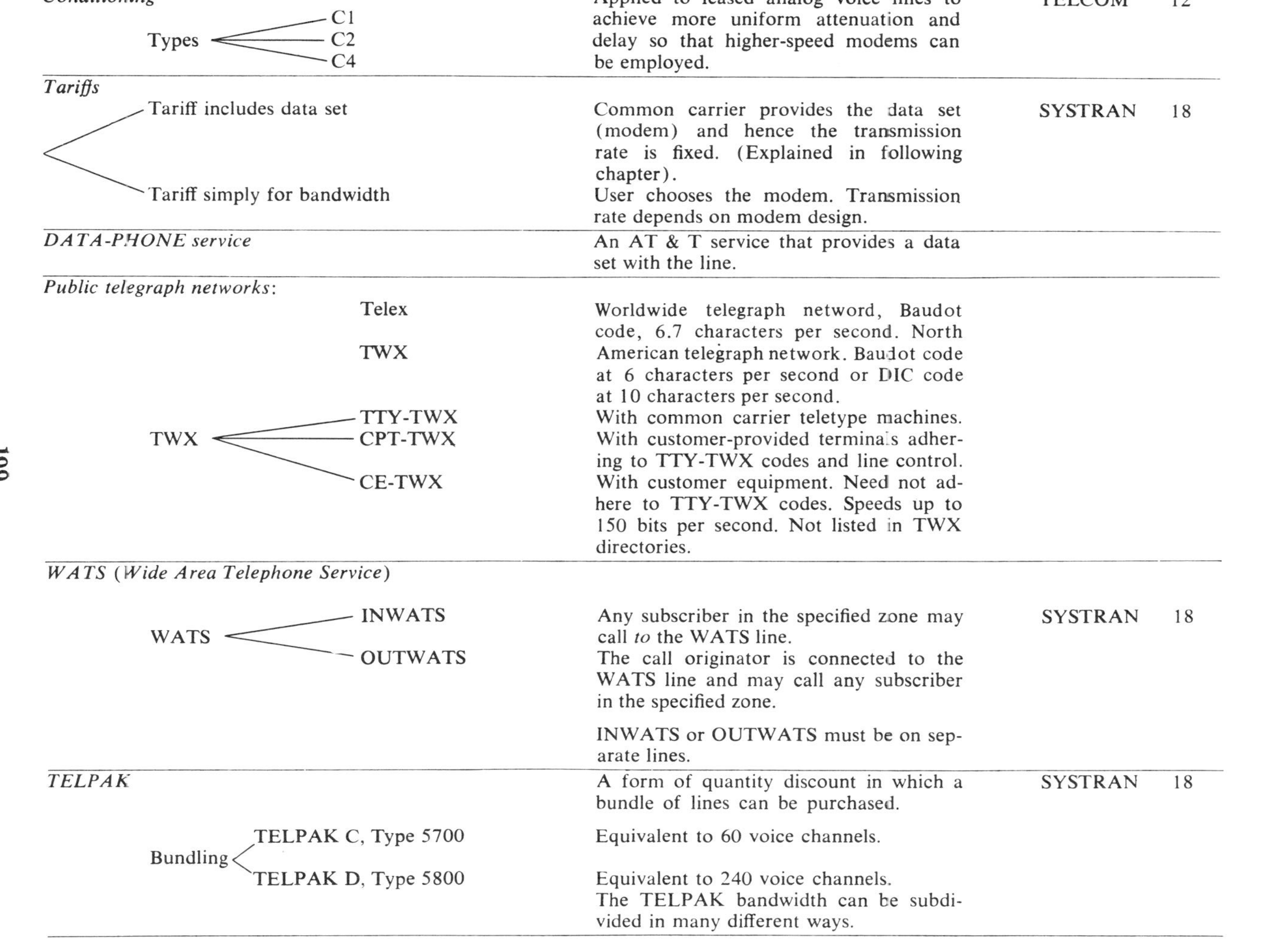

Term	Description	Glossary	Chapter
Conditioning Types: C1, C2, C4	Applied to leased analog voice lines to achieve more uniform attenuation and delay so that higher-speed modems can be employed.	TELCOM	12
Tariffs			
Tariff includes data set	Common carrier provides the data set (modem) and hence the transmission rate is fixed. (Explained in following chapter).	SYSTRAN	18
Tariff simply for bandwidth	User chooses the modem. Transmission rate depends on modem design.		
DATA-PHONE service	An AT & T service that provides a data set with the line.		
Public telegraph networks:			
Telex	Worldwide telegraph netword, Baudot code, 6.7 characters per second.		
TWX	North American telegraph network. Baudot code at 6 characters per second or DIC code at 10 characters per second.		
TWX: TTY-TWX	With common carrier teletype machines.		
TWX: CPT-TWX	With customer-provided terminals adhering to TTY-TWX codes and line control.		
TWX: CE-TWX	With customer equipment. Need not adhere to TTY-TWX codes. Speeds up to 150 bits per second. Not listed in TWX directories.		
WATS (*Wide Area Telephone Service*)			
WATS: INWATS	Any subscriber in the specified zone may call *to* the WATS line.	SYSTRAN	18
WATS: OUTWATS	The call originator is connected to the WATS line and may call any subscriber in the specified zone.		
	INWATS or OUTWATS must be on separate lines.		
TELPAK	A form of quantity discount in which a bundle of lines can be purchased.	SYSTRAN	18
Bundling: TELPAK C, Type 5700	Equivalent to 60 voice channels.		
Bundling: TELPAK D, Type 5800	Equivalent to 240 voice channels. The TELPAK bandwidth can be subdivided in many different ways.		

TYPES OF COMMUNICATION LINES AVAILABLE

1. *Public* (*Dial-up*) *Lines*	Bit Rate (bits per second): Fixed	Bit Rate (bits per second): Dependent on Modem	Bandwidth (k Hz)	Type of Line: United States: AT & T	Type of Line: United States: Western Union	Type of Line: United Kingdom	Half Duplex or Full Duplex
Subvoice grade	45				TTY-TWX and CPT-TWX		HDX
		Up to 45			CE-TWX		HDX or FDX
	50				Telex	Telex	HDX
	110				TTY-TWX and CPT-TWX		HDX
		Up to 150			CE-TWX		HDX or FDX
		Up to 200				Datel 200	FDX
The public telephone network		Up to 600	3	Public network		Datel 600	FDX
		Up to 1200	(Not all freely usable because of network signaling.)			Datel 600	FDX
		600 to 9600					HDX or FDX
Switched wideband networks	600 1200 & 2400 4800* 9600* 38,400*	(Other speeds will be achievable with other modems.)	2 4 8* 16* 48*		BEX (certain cities only)		FDX
	50,000	Up to 50,000		Dataphone 50 (few cities only)			FDX

2. *Leased Lines*

Subvoice grade		Up to 45		1004			HDX/FDX
	50					Tariff H	FDX
		Up to 55		1002			HDX/FDX
		Up to 75		1005			HDX/FDX
	100					Datel 100	FDX
		Up to 150		1006			HDX/FDX
		Up to 180			1006		HDX/FDX
	200					Datel 200	FDX
Voice-grade lines (For the higher rates, conditioning is needed.)		Up to 600	3			Datel 600	FDX
		Up to 1200	3			Datel 600	FDX
		600 to 10,500	3	3002 (C1, C2, and C4 conditioning)	3002	(Datel 2000 refers to conditioned or high-quality voice lines)	HDX/FDX
Wideband	19,200		24	8803	8803		FDX
	40,800		48	8801	8801	Special quotation	FDX
	50,000		48	8801	8801		
	230,400		240	5700 or [Originally called TELPAK C and D (see text), can be used as "bundles" of small bandwidth lines.]	5800		FDX

*Planned but not yet available.
Book: SYSTRAN, Chapters 17 and 18.

D TRANSMISSION TECHNIQUES

1. Modes of Transmission

DISADVANTAGES OF THE VARIOUS TRANSMISSION MODES

MODE OF TRANSMISSION		ADVANTAGES AND DISADVANTAGES
Four-wire		Permits full-duplex transmission.
Two-wire		Full-duplex transmission still possible with separate frequency bands for the two directions.
Simplex		Rarely used for data transmission, as there is no return path for control, or error signals.
Half duplex		Commonly used for data transmission, though a full-duplex line may cost little more. (Often 10% more in the United States.)
Full duplex	Data in both directions at once	System sometimes cannot take advantage of this, as data cannot be made available for transmitting in both directions simultaneously. Can substantially reduce the response time, however, on a conversational multidrop line. Often requires a more expensive terminal. Commonly used on a link between concentrator and computer.
	Data in one direction; control information in the other	A common arrangement, though, as data are still only being sent in one direction at a time, half-duplex transmission may give better value for money at low character rates. With high character rates the line turnaround time may be long compared to the character time and full-duplex operation may eliminate most turnaround delay.

Mode of Transmission		Advantages and Disadvantages
Serial-by-character Parallel-by-bit	Separate wires	Low transmitter cost, but high line cost. Economical for in-plant use. Line costs too expensive for long distances.
Serial-by-character Parallel-by-bit	Separate frequencies	Used on voice lines to give a slow but inexpensive terminal. For efficient line utilization, however, data set costs are high, and receiver cost can be high.
Serial-by-character Serial-by-bit		The most common system, especially on long lines.
	Start-stop transmission	Inexpensive terminal, e.g., telegraph machines. Only one character lost if synchronization fails. Not too resilient to distortion at high speeds.
	Synchronous transmission	More expensive terminal. Block lost if synchronization fails. Efficient line utilization. High ratio of data to control bits. More resilient to noise and jitter than start-stop transmission, especially at high transmission speeds. The most common system on lines of 600 bits per second and faster.
	High-speed pulse train	In-plant or private wiring only at present. Low wiring cost with low terminal cost. High accuracy.

2. Techniques for Controlling Line Errors

Technique	Usage	Book	Chapter
1. Ignore line errors.	Used where proportion of operator errors is far greater than line errors. Used in verbal message transmission.	SYSTRAN	9
2. Common error-detection codes. — Character coding (e.g., 4-out-of-8 code)	Less commonly used today. Less effective than block codes.	NETWORK	2
— Parity checks	Commonly used but of little value (especially with higher-speed transmission).	SYSTRAN	15
— Vertical and longitudinal checks	Commonly used. Simple circuitry. Less efficient than polynomial codes.	SYSTRAN	15
— Polynomial checks on blocks; Checks on groups of records	Exceedingly efficient block error detection is possible with polynomial codes.	NETWORK	5
3. Error detection without automatic retransmission.	Manually initiated retransmission is used when the proportion of operator errors is high (e.g., on airline reservation systems).	SYSTRAN	15
4. Error detection with automatic retransmission.	The safest method, given a good error-detecting code. Some form of message storage is needed at the terminal.	SYSTRAN	15
— Retransmit one character	Echo check.		
— Retransmit one word	Word check.		
— Retransmit one message or record	Probably the most common. Block check on the message or record.		
— Retransmit several messages or records	Possibly checked with a high-order polynomial code.		
— Retransmit a batch	Possibly checked with a batch hash total.		
Note:	It pays to select optimum block size or batch size.	SYSTRAN	35

Technique	Usage	Book	Chapter
5. Retransmit at a later time.	For example, after a periodic or daily balancing or policing run.	SYSTRAN	15
6. Retransmit several times if first attempt fails.	Normally done.	SYSTRAN	15
7. Retransmit several times with variable time delays.	To circumvent lengthy periods of line noise.	SYSTRAN	15
8. Retransmit with modem switched to lower speed.	Can be done automatically on certain terminals with built-in modems. Otherwise the modem speed may be switched manually.	SYSTRAN	14
9. Loop check. — Characters returned to terminal — Characters sent twice by terminal	Simple where loop transmission is employed. Used on in-plant loops with capacity to spare.	SYSTRAN	15
10. Forward-error correction.	Less efficient and less safe than error detection. Used on one-way links and on links with abnormally high error rates such as long-wave and short-wave radio, and telephone-line modems operating at 9600 bits per second.	SYSTRAN	15
11. Combination of forward-error correction and error detection with retransmission.	Forward-error correction is added to error-detection schemes when the proportion of retransmission becomes high enough to degrade throughput significantly.	SYSTRAN	35

3. Criteria for Choice of Modem*

Basic criteria:

1. *Cost.* Cost of modem, or overall cost per bit transmitted.
2. *Speed.* Bits per second.
3. *Error rate.* Fraction of bits incorrect. Resistance to noise bursts.

Other important criteria:

1. *Reliability.* Mean time between failures.
2. *Response time.* What is the modem's turnaround time? How long does the establishment of carrier take when switching terminals on a polled line? Do either of these factors harm the response time?
3. *Start-stop transmission*? Can the modem handle start-stop and synchronous transmission?
4. *Network compatibility.* Is the modem compatible with the network signaling? In a country with government-controlled communication, does it have approval?
5. *Alternate speeds.* Can it transmit at different speeds with different degrees of error protection? Can it be switched between private or public lines?
6. *Portable*? Should the terminal and modem be portable?
7. *Conditioning*? Does it need the line to be conditioned?

*Book: SYSTRAN, Chapter 14.

4. Modems: Types of Devices

Technique	Cost	Speed	Errors	Advantages and Disadvantages
	L = low M = medium H = high			
1. No modems	L	L	L	Short distance only. Wire pairs without amplifiers. Generally suitable for low-speed operation only (< 300 bits per second) except over lines of only a few thousand feet. The cost of the modem is saved.
2. Acoustical coupler	L	L	M	Portable. No wired connection to the telephone line, hence convenient. Generally suitable for low-speed operation only < 300 bits per second).
3. Multifrequency transmission (like acoustical coupler or Touch-Tone telephone)	L	L	M	Inexpensive. Generally not suitable for high-speed operation.
4. Modulation of a sine wave carrier — Amplitude modulation (AM)	L	M/H		Generally the least expensive of the three. The worst affected by impulse noise.
4. Modulation of a sine wave carrier — Frequency modulation (FM)	M	M/H		Less affected by noise than amplitude modulation. Less affected by delay distortion than phase modulation.
4. Modulation of a sine wave carrier — Phase modulation (PM)	M	M/H		Generally the most expensive. Generally the least affected by noise. Usually employed on conditional lines. Tends to have the longest modem turnaround time, or start-up time.
5. Multilevel transmission — Number of states: 2 4 8	L M M		L M H	2 states are used to send bits; 4 states for di-bits; 8 states for tri-bits. 8 states give three times the speed of 2 states but with much higher number of of errors

Technique	Cost	Speed	Errors	Advantages and Disadvantages
	L = low M = medium H = high			
6. Different detection methods				See table on following page.
7. Conditioned lines	M			Required for higher-speed modems, especially with phase modulation. Increases line cost. Only on private lines.
8. Dynamic equalization	H	H		A form of automatic variable conditioning for modems on dial-up lines. Expensive. Permits high-speed operation.
9. Randomizing signal in modem	M			Used to overcome problems with signaling frequencies on dial-up lines. Digital circuitry in modem. Expensive. Permits higher-speed operation on public lines.
10. Forward-error correction	H	H	L	Used in high-speed modems to combat the high error rates. Expensive. Increases turnaround time. Required digital circuitry in modem.

Book describing modulation techniques: TELCOM, Chapter 13.
Book discussing advantages and disadvantages: SYSTRAN, Chapter 14.

5. Modems

THEORETICAL PERFORMANCE OF DIFFERENT MODULATION SYSTEMS IN THE PRESENCE OF WHITE NOISE*

System	Number of States	Speed in Bits per Second per Cycle of Bandwidth	Average Signal-to Noise Ratio (dB) for Error Rate of 1 in 10^4 Bits
Polar baseband	2	2	11.4
	4	4	18.3
	8	6	24.3
	16	8	30.2
Full carrier Amplitude Modulation:			
envelope detection	2	1	11.9
coherent detection	2	1	11.4
	4	2	19.8
	8	3	26.5
Suppressed carrier Amplitude Modulation,			
coherent detection	2	1	8.4
	4	2	15.3
	8	3	21.3
	16	4	27.2
Frequency Modulation	2	1	11.7
	4	2	21.1
	8	3	28.3
Phase Modulation:			
coherent detection	2	1	8.4
	4	2	11.4
	8	3	16.5
	16	4	22.1
	32	5	28.1
Phase Modulation:	64	6	34.1
differential detection	2	1	9.3
	4	2	13.7
	8	3	19.5

**Data Transmission*, by Bennett and Davey, McGraw-Hill, 1965.

E NETWORK STRUCTURES

1. The Attachment of the Communication Lines to the Main Computer

Method	Advantages and Disadvantages	Book	Chapter
Method 1: The communication lines go directly into the main computer. (They may bypass the I/O channels, as with the "integrated line attachment" on the IBM 360 Model 25.)	Advantages: Reduces the number of units needed (and their floor space), and hence the cost. Duplicate multiplexing (in the line-control unit and then in its channel) is avoided. Disadvantages: A large number of communication line interrupts must be serviced in the main computer. The method tends, therefore, to be used on small systems with small numbers of lines.	DESREAL	19
Method 2: A line adaptor is attached to the computer I/O channels. The quantity of data stored and sent at one time on the computer channel can be: One byte / A group of bytes / One message	Advantages: Often less computer interrupts than method 1 above. The larger the quantity of data stored and sent to the computer at once, the fewer the number of interrupts. A variety of different adaptors can be available for different types of transmission. Disadvantages: Much is left to the main computer software. This can result in inefficient use of the main computer, especially on systems with many lines and diverse control procedures.		
Method 3: The lines terminate in a separate line-control computer which relieves the main computer of line and network control functions.	Advantages: The work of communication-line handling is removed from the main computer, thus saving processor time and core. Line-control software is not constrained by operating system requirements. Flexible; special line and terminal requirements can be accommodated	SYSTRAN	26

Method	Advantages and Disadvantages	Book	Chapter
	in the software. The same machine can handle a wide range of line types and terminal types. The machine can change its capability when new devices are dialed. The line-control computer can preprocess the messages. It can edit them, carry out validity checks, take hash totals, etc. It may store responses and, in some cases, may itself respond to the terminals. It may handle voice answerback responses. Because it is an extra processor, it increases the possibilities for graceful degradation when failures occur. Software failure in the main computer does not knock down the network. A new IPL (initial program load) can take place in the main computer without affecting the line control computer or network. One device can have a wide range of capabilities. This increases the potentialities for system enlargement and development. Disadvantages: More expensive for small systems (though less so for large ones). Possibly lowers system reliability because two processors are in series. If made by a different manufacturer to the main processor may introduce problems of interfacing, modification, or maintenance.		

2. Leased-Line Network Structures

		Advantages and Disadvantages	Book	Chapter
1. Simple point-to-point lines	Advantages:	Simple. No queuing. No busy signals.	NETWORK	6
	Disadvantages:	Expensive except for very short distance connections. The techniques below which lower the total line cost are of greater value with longer lines.		
2. Multipoint "tree-structure" lines	Advantages:	Line-cost saving which becomes large on long-distance networks.	NETWORK	9
	Disadvantages:	Queuing or contention problems and hence increased response times. Line-control logic needed at terminals. Logic and software needed at computer.		
Line control — Contention	Advantages:	Minimal logic requirements.		
	Disadvantages:	User can be kept waiting.		
Line control — Polling — Roll call	Advantages:	Less logic required at drop points than with hub polling.		
	Disadvantages:	Many more line turnarounds than with hub polling, hence much longer response times when a modem giving a long turnaround time is used (as with modern high speed modems).		
Line control — Polling — Hub	Advantages:	Few line turnarounds, hence better response times.		
	Disadvantages:	More logic required at drop points, including that to cover drop point failure.		

	Advantages and Disadvantages	Book	Chapter
3. Use of private exchanges and point-to-point leased lines	Advantages: Simple. No queuing, hence no problem caused with response times. Disadvantages: Some users will obtain "busy" signals. Access to computer is not guaranteed. Line cost required to achieve low probability of busy signals may be higher than with some other network structures.	**NETWORK**	8
4. Multiplexers (which combine the signals from a number of devices onto a smaller number of physical transmission paths (often one) in such a way that *every device has a subchannel available to it.*)	Advantages: Simple, inexpensive, generally reliable devices. No change in programming necessary; the multiplexer can be "transparent" to the programs. No significant increase in response time. Multiplexers cause no contention or queuing problems and no "busy" signals. Disadvantages: Some other schemes may give a lower overall network cost.	**NETWORK**	11
— Frequency-division multiplexing	Can be built like a modem so that different "drops" can be connected to one line at different locations.		
— Time-division multiplexing	Use digital logic circuitry which is fast dropping in cost, and hence they are becoming less expensive than frequency-division multiplexing devices, modems can be avoided between the terminal and the multiplexing point.		
5. Shared buffers (organized to take advantage of high-speed transmission between the buffer and the computer)	Advantages: Permits synchronous transmission with low-cost asynchronous terminals. Permits efficient line usage. Cheaper than putting the buffers in the individual terminals.	**NETWORK**	7

	Advantages and Disadvantages	Book	Chapter
6. Multiterminal control units	Advantages: Same as shared buffers, with the added advantage that logic for terminal control is built into the device (error control, polling logic, screen regeneration, etc.). With editing features in the control unit, the number of characters transmitted on the line can be minimized.	NETWORK	10
7. Concentrators (Concentrators combine the signals from a number of devices onto a smaller number of physical transmission paths (often one) in such a way that there is a certain probability that a device will have a subchannel available to it. Unlike a multiplexer, the concentrator does not provide a permanent subchannel for each device.)	Advantages: Can give more efficient use of the high-speed lines and hence lower network cost than simple multiplexers. Can give better error control than simple multiplexers. Fast response time possible. No "busy" signals. Disadvantages: More expensive than simple multiplexers. Cost trade-off depends on line configuration and needs a detailed set of calculations. Reliability may be lower than with simple multiplexers. Unlike multiplexers, software is usually required at the computer.	NETWORK SYSTRAN	12 42
8. Concentrators with multidrop low-speed lines	Advantages: Possible lower-cost network. Disadvantages: More complex because the low-speed lines must be polled. More to go wrong. Response time is degraded, often severely. by queuing on the low-speed lines.	NETWORK SYSTRAN	12 43

	Advantages and Disadvantages	Book	Chapter
9. Concentrators on multidrop high-speed lines	Advantages: Possible lower-cost network. Because the multidrop lines are high speed, the degradation in response time is much less severe than when low-speed multidrop lines are used. May be less expensive than a concentrator, which has to poll the low-speed lines. Disadvantages: More complex than some of the previous schemes.	**NETWORK** **SYSTRAN**	12 43
10. Stored-program computers used as concentrators	Advantages: Flexible. Compaction techniques can substantially reduce the number of characters flowing on the lines, and hence the network cost. May be programmed to give a limited intelligent response to the terminal in the event of the central computer failing. Incompatible terminals may be handled. Disadvantages: Expensive. Failure is more likely than with simpler machines. Increased programming requirement. Increased maintenance requirements.	**NETWORK**	13
11. Stored-program concentrator computer with application programs	Advantages: Much of the transmission to the central computer is eliminated, and may occur only when the power or the data base of the central machine is needed.	**NETWORK**	13 & 15

	Advantages and Disadvantages	Book	Chapter
	Disadvantages: The expense may not justify the line-cost savings. Application programs not all in one location. Careful control needed when programs are modified. Reliability may be degraded (the central computer may be duplexed, but duplexing the peripheral computers is usually too expensive).		
12. Message-switching systems	Advantages: Inexpensive line network. System stores the messages. If a location cannot receive a message, the system delivers it later when it can. Broadcasting, multiaddress messages, and automatic distribution lists are handled. Disadvantages: Traditionally designed without fast response-time requirements. Is economic only above a certain traffic volume.	NETWORK	14
13. Packet-switching systems (designed for routing a preformatted data packet at high speed through a multinode network)	Advantages: Can provide a switched network between incompatible terminals. No blocking (busy signals)—unlike line switching. Can give very fast transit times and very efficient utilization of high-speed trunks. Alternate routing provides protection from node or trunk failure. Disadvantages: Economic only above a certain high traffic volume. Cost compari-	SYSTRAN	21

	Advantages and Disadvantages	Book	Chapter
	son with line-switching network needs careful evaluation. The interface with the terminals may restrict certain types of devices (unlike a line-switched network).		
14. Looped lines between terminals or concentrators	Advantages: Can give protection from line failure if a mechanism is incorporated for accessing devices on the looped line when a break occurs. Can give very fast response times because no line turnarounds are needed (when the loop is intact). Highest-speed modems can be used because there are no line turnarounds. Very high bit rates are possible using PCM (only used on in-plant system at time of writing). Disadvantages: Give a higher line mileage than with tree-structure lines. Multiplexing logic is needed at each drop point.	**SYSTRAN**	21
15. Distributed networks (giving an alternate path to each terminal or concentrator)	Advantages: Protection from leased-line failures. Disadvantages: Higher network cost than tree-structure lines. Facilities required to switch to the alternate line configurations when failures occur and to alternate polling lists.	**SYSTRAN**	21

3. Basic Multiplexer Features. A Checklist*

Multiplexing refers here to a device that combines several data signals to travel over one higher-speed line, but does not restructure or process the data. The following pages discuss concentration devices that modify of process the data or carry out computing functions.

1. What low-speed line speeds or terminal speeds can be used?
2. How many low-speed transmissions can be multiplexed?
3. Can it handle terminals with different speeds?
4. Can terminals operating with different speeds be intermixed at one time? If so, in what groupings?
5. Can terminals operating with different codes be used or intermixed?
6. Data transparency. Is the device capable of passing all combinations of bits?
7. Can the device pass all necessary control signals?
8. What speed is the higher-speed line? 9600 bits per second on a voice line?
9. Can different speeds be used for the higher-speed line?
10. Must the higher-speed line be full duplex or half duplex?
11. Can some terminals transmit while others receive?
12. Does the device carry out modulation or are additional modems needed?
13. Must the terminals all be at one multiplexer location or the low-speed lines all come to all location, or can the high-speed line wander round many terminal locations–a multiplexor star configuration or a terminal string. (Book: NETWORK, Figure 11.2 and text.)
14. Has the device contention ability? In other words, can the number of low-speed lines be greater than the number that can operate simultaneously?
15. Does it have good error-control techniques on the high-speed line?
16. Does it have good error-control techniques on the low-speed line?
17. When an error occurs on the high-speed line, does retransmission take place automatically for the multiplexer, or must retransmission be from the terminal?
18. Is the device of high reliability? Does its manufacturer have a history of products with proven reliability?
19. Is it easily maintainable and does the price include maintenance?
20. Does it require another multiplexer at the computer interface; if so, are they completely transparent to the programming?
21. Alternatively can its high-speed line go directly into the computer so that software demultiplexing is used? Which is the cheaper?
22. Can its capacity be expanded easily?
23. Does its price and capability give a minimum-cost network for the sys-

*Book: SYSTRAN, Chapter 20.

tem in question, when compared with other devices? (See calculations in SYSTRAN, Chapter 42).

24. Can multidrop lines be used downstream of the device?

4. Features of "Intelligent" Concentrators or Control Units. A Checklist*

The simplest form of concentrator is a contention device that links a number of lines from terminals (or telephones) to a smaller number of similar lines to the computer (or switching office). Here we list features of concentration devices with logic features for manipulating the data being transmitted.

1. Is the machine a stored-program device?
2. If so, is the necessary software for it available?
3. Is such software well documented, modular, and easily modified?
4. Can the terminals attached be low-cost, asynchronous, unbuffered devices?
5. Can the line to the computer be synchronous, and is it organized as efficiently as possible?
6. Can a mixture of different terminals be attached?
7. Can these have different codes, line speeds, and line-control techniques?
8. What speed lines can be used?
9. How many low-speed lines can be used?
10. Can terminals be multidropped on the low-speed lines?
11. If so, does the concentration device handle all the polling on the low–speed lines?
12. What high-speed lines can be used? Wideband?
13. Can different high-speed lines be used?
14. Can several concentration devices be attached to one high-speed line?
15. Is the high-speed polled? Is hub polling used?
16. Data transparency. Is the device capable of passing all combinations of bits?
17. Must the higher-speed line be full duplex or half duplex?
18. Can some terminals transmit while others receive?
19. Does the device carry out modulation or are additional modems needed?
20. How many low-speed lines can be operating simultaneously? What is the throughput of the device?
21. Does it have good error-control techniques on the high-speed line?
22. Does it have good error-control on the low-speed line?
23. When an error occurs on the high-speed line, does retransmission take place automatically for the device?

*Books: NETWORK, Chapter 15 and SYSTRAN, Chapter 26.

24. Is the device of high reliability? Does its manufacturer have a history of products with proven reliability?
25. Is it easily maintainable and does the price include maintenance?
26. Does it require special hardware at the computer interface?
27. Does it require special software in the computer? Is this available, modular, well documented, and easily modified?
28. Can its capacity be easily expanded?
29. Does it have features for maintaining security?
30. What action does it take when line failures occur? Can it switch to an alternate line or an alternate dialed connection?
31. Can it carry out message-editing features if this is not done in the terminal?
32. Are any data-compaction techniques used? If so, how powerful are they?
33. Can the machine generate responses to the terminal on receipt of brief commands from the central computer?
34. Can the machine check operator input—for example, syntax checks, reasonableness checks, validity checks?
35. Can the machine generate error responses to the terminal?
36. Can the machine carry on simple dialogue with the terminal, as in the collection of a set of input data, and thus lessen the quantity of data transmitted to and fro?
37. Can the machine carry out limited computing functions? (See illustration of a banking system in NETWORK, Chapter 13).
38. Does the machine have direct-access files?
39. Does its price and capability give a minimum-cost network for the system in question, when compared with other devices? (See calculations in SYSTRAN, Chapter 42.)

F SOFTWARE

1. SUMMARY OF FEATURES IN SOFTWARE FOR DATA TRANSMISSION*

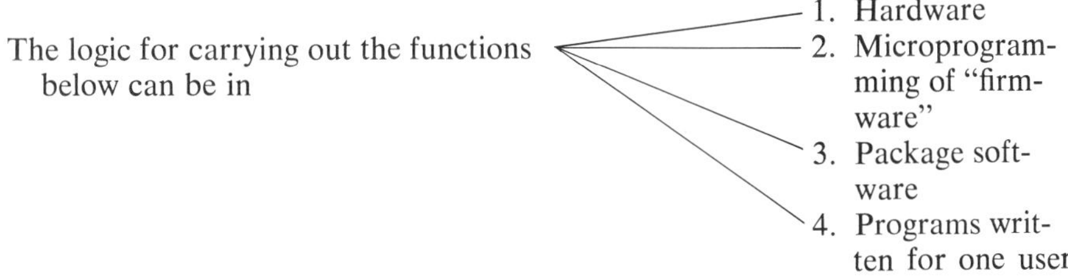

A. Logic Requirements for Basic Transmission Functions

On reception:

1. To initiate and control the reception of data.
2. To assemble the bits into characters.
3. To assemble the characters into messages.
4. To recognize end-of-record and end-of-transmission characters.
5. To convert the coding of characters.
6. To chcck for crrors.
7. To correct errors or to initiatc retransmission of erroneous messages.
8. To edit incoming messages, removing backspace and other such characters.
9. To deliver messages to the main programs.

*Book: SYSTRAN, Section V.

On transmission:

10. To accept messages from the main programs.
11. To prepare them for transmission, adding appropriate control and address characters.
12. To initiate the transmission.
13. To terminate the transmission when the end-of-transmission character is sent.
14. To monitor the transmission process.
15. To accept acknowledgement from the receiving terminal.
16. To retransmit messages that were received with errors or not received.

B. For Directing Transmission to the Relevant Device

For dial systems:

17. Dialing a remote device.
18. Scanning dial-up terminals.
19. Automatically answering when the computer is dialed.
20. Initiating transmission after a dialed connection.
21. Determining the type of device that has dialed in.

For polling systems:

22. Polling terminals to see whether any of them have something to transmit.
23. Receiving the terminal response and initiating reception.
24. Sending a message to a terminal asking permission to transmit.
25. Receiving the terminal response and initiating transmission.
26. Resetting the line after transmission.
27. Time-out functions if a device fails to respond.
28. Modifying the polling list when a device fails or is removed.
29. Changing the forwarding address of a device when hub polling is used, if the terminal at that address fails.
30. Recovery from addressing errors.

For loop systems:

31. Establishing and maintaining loop synchronization.
32. Placing the characters into appropriate transmission slots.
33. Receiving characters from appropriate transmission slots.
34. Detecting errors on reception and instructing the device to retransmit.
35. Accepting notification of outgoing errors, and retransmitting.

C. *Blocking, Multiplexing, and Compacting*

36. Combining several "logical" messages into one "physical" transmission message.*
37. Extracting the "logical" messages on reception.
38. Where a remote multiplexer is used, the multiplexing and demultiplexing operation at the computer center may be done by software.
39. Data transmitted may be compacted by a variety of possible techniques for expansion at a remote device.
40. References to precomposed messages may be sent and these messages generated at a remote device.
41. Decompaction may take place on input.

D. *Scheduling and Resource Allocation*

42. Buffer allocation. Dynamic allocation of buffers is employed to give efficient storage utilization.
43. Handling queues for lines.
44. Handling queues for programs.
45. Handling items for multiple destination.
    Priority scheduling (e.g., giving priority to certain message types).
46. Store and forward functions.
47. Message switching (see list of functions on following pages).

E. *Logging and Statistics Gathering*

48. Message logging (for later reference and for recovery procedures).
49. Keeping statistics of traffic volumes.
50. Keeping statistics of line errors.
51. Keeping statistics of line failures.

F. *Control of an External Answering Unit*

52. Control of an external unit which generates responses.
53. Control of an external unit which formats documents and displays.
54. Control of an external voice answerback unit.

G. *Handling Failures*

55. Generation of message-sequence numbers for use in message-recovery procedures.

*Book: SYSTRAN, Chapter 27.

56. Recovery routines associated with sequential numbering.
57. Recovery routines associated with message logging.
58. Initiating dial-up when a leased line fails.
59. Hash totaling of incoming transactions.
60. Responding to terminals when there is a partial failure of the central system.

H. Security Functions

61. Message encryption and decryption.
62. Identification of the terminal sending input.
63. Identification of the terminal before sending output.
64. Identification of the terminal user—for example, with security codes.

I. Diagnostics

65. Diagnostics for checking correct operation of a line and terminal from a remote terminal.
66. Cross-patching tests to ensure that the hardware and software of the line-control computer are working correctly.
67. Software for sending and automatically receiving a diagnostic response from a remote device.
68. Determination of which devices are inoperative on multidevice lines.
69. Maintenance of network and device status tables.

J. Dialogue Functions. (These are mostly application oriented.)

70. Reasonableness checks on input.
71. Completeness checks on input.
72. Context editing of input.
73. Page-flip, scroll, index searching, browsing, or other operations for viewing but not processing data.
74. Dialogues with the operator preliminary to the use of application programs (e.g., for data collection).

2. CHECKLIST OF FUNCTIONS THAT CAN BE PERFORMED BY MESSAGE-SWITCHING SOFTWARE*

1. The software accepts messages from distant terminals. The terminals are often teleprinters and paper-tape readers, but other devices may be used, such as card readers and special input keyboards. The system may also accept messages from other computers.
2. On receipt of a message, it analyzes the message's header to determine the destination or destinations to which the message must be sent.
3. The system may analyze the header for a priority indication. This will tell the program that certain messages are urgent. They must jump any queues of messages and be sent to their destination immediately.
4. It may analyze the header for an indication that some processing of the message is necessary; for example, statistical information from the message may be gathered by the system.
5. The system detects any errors in transmission of the incoming message and requests a retransmission of faulty messages. This retransmission may be automatic.
6. It detects format errors in incoming messages as far as possible. Types of format, errors that may be picked up include the following: (1) Address invalid. The address to which the message is to be sent is not included in the computer's directory. (2) Excessive addresses. There are more than the given maximum number of addresses allowed. (3) Incorrect format. An invalid character, for example, the control character appears in the message in an incorrect location. (4) A priority indicator is invalid. (5) Originator code error. The address of the originator is not included in the computer's list. (6) Incorrect character counts.
7. The system stores all the messages arriving and protects them from possible subsequent damage.
8. It takes messages from one store and transmits them to the desired addresses. One message may be sent to many different addresses. In doing this, it does not destroy the message held in the store. The store is thus a queuing area for messages received and messages waiting to be sent, as well as a file in which messages are retained.
9. The system redirects messages from the store and sends them to the terminals requesting them. It may, for example, be asked to resend all messages from a given serial number or to resend a message with a specified serial number.
10. Systems in use store messages in this manner for several hours or, on some systems, several days. Any message in the store is immediately accessible for this period of time.
11. The system may also maintain a permanent log of messages received. This will probably be done on a relatively inexpensive medium, such as magnetic tape, and not on a random-access file.

*Book: NETWORK, Chapter 14.

12. If messages are sent to a destination at which the terminal is temporarily inoperative, the system intercepts these messages. It may automatically reroute them to alternative terminals that are operative. On the other hand, it may store them until the inoperative terminal is working again.
13. It may intercept messages for other reasons. For example, the system may be programmed to send a message to the location of an important person, although he may be moving from one place to another. The person in question leaves his current location address with the computer, and the computer diverts messages for him to that location.
14. The system maintains an awareness of the status of lines and terminals where possible to make a log of excessive noise on lines, and to notify its operator when a line goes out. The system maintains records of any faults it detects.
15. The message may be given serial numbers by the operator sending them. The computer checks the serial number and places new serial numbers on the outgoing messages. When serial numbers are used, the system can be designed to avoid the loss of any message.
16. At given intervals, perhaps once an hour, the system may send a message to each terminal, quoting the serial number of the last message it received from that terminal. The terminal's operator then knows that the switching system is still on the air.
17. The system may conduct a statistical analysis of the traffic that it is handling.
18. It may be programmed to bill the users for the messages sent. It may, for example make a small charge per character sent from each terminal and bill the terminal location appropriately.
19. It produces periodic reports of its operation for its operator. These may include reports on the status of all facilities, error statistics, reports giving the number of messages in each queue, message counts, and so on.

REFERENCES

The references each give that chapter or section in the other books of this set in which further reading can best continue. All the books are by James Martin, published by Prentice-Hall Inc., Englewood Cliffs, N.J.

	REFERENCE		
Chapter 1	1	The Computerized Society	
Chapter 2	1	Systems Analysis for Data Transmission	Chapter 7
Chapter 3	1	Telecommunications and the Computer	Chapter 8
	2	Telecommunications and the Computer	Chapter 10
	3	Telecommunications and the Computer	Chapter 11
	4	Future Developments in Telecommunications	Chapter 6
	5	Telecommunications and the Computer	Chapter 15
Chapter 4	1	Telecommunications and the Computer	Chapter 8
	2	Telecommunications and the Computer	Chapter 17
	3	Telecommunications and the Computer	Chapter 12
	4	Telecommunications and the Computer	Chapter 9
Chapter 5	1	Telecommunications and the Computer	Chapter 13
	2	Telecommunications and the Computer	Chapter 17
	3	Systems Analysis for Data Transmission	Chapter 14
Chapter 6	1	Systems Analysis for Data Transmission	Chapter 16
Chapter 8	1	Design for Man-Computer Dialogues	Chapter 27
	2	Systems Analysis for Data Transmission	Chapter 35
Chapter 9	1	Design of Man-Computer Dialogues	Chapters 6 and 7
		Systems Analysis for Data Transmission	Chapter 10
Chapter 10	1	Telecommunications and the Computer	Chapter 8
	2	Future Developments in Telecommunications	Chapter 17
	1	Systems Analysis for Data Transmission	Section VI
	2	Telecommunications and the Computer	Chapter 18
	3	Future Developments in Telecommunications	Chapter 19
	4	Systems Analysis for Data Transmission	Chapter 33
	5	Teleprocessing Network Organization	Chapter 14
	6	Systems Analysis for Data Transmission	Section VI

	REFERENCE		
Chapter 12	1	Teleprocessing Network Organization	Chapter 9
Chapter 13	1	Design of Man-Computer Dialogues	
	2	Systems Analysis for Data Transmission	Chapter 7
Chapter 14	1	Systems Analysis for Data Transmission	Chapter 23
	2	Security, Accuracy, and Privacy in Computer Systems	
	3	Design of the Man-Computer Dialogues	
Chapter 15	1	Programming Real-Time Computer Systems	Chapter 16
Chapter 16	1	Teleprocessing Network Organization	
Chapter 17	1	Systems Analysis for Data Transmission	Chapter 7
	2	Systems Analysis for Data Transmission	Chapter 36
	3	Design of Real-Time Computer Systems	Chapter 24
	4	Systems Analysis for Data Transmission	Chapter 7
	5	Systems Analysis for Data Transmission	Chapter 32
	6	Systems Analysis for Data Transmission	Chapter 33
	7	Design of the Man-Computer Dialogues	
	8	Teleprocessing Network Organization	Chapter 7
	9	Teleprocessing Network Organization	Chapter 9
	10	Systems Analysis for Data Transmission	Chapter 14
	11	Systems Analysis for Data Transmission	Chapter 2
	12	Systems Analysis for Data Transmission	Chapter 39
	13	Systems Analysis for Data Transmission	Chapter 40
	14	Systems Analysis for Data Transmission	Chapter 41
	15	Systems Analysis for Data Transmission	Chapter 44

CLASS QUESTIONS

The questions labeled with an asterisk are particularly suited to class discussion. The author's book *Systems Analysis for Data Transmission* contains more questions. For advanced class discussions, the instructor might try Part VII of the questions in that book—"Thought Provokers."

1. What is the difference between "bauds" and "bits-per-second"?
2. What are simplex, half-duplex, and full-duplex lines?

*3. A full-duplex line in the United States is commonly 10 percent more expensive than a half-duplex line. What facilities should the terminals have to take *maximum* advantage of full-duplex transmission? When is the expense of such facilities warranted?

4. What does modulation mean?
5. What are the common methods of modulation?
6. What is the difference between synchronous and asynchronous transmission? Draw a typical message format for each. What are the relative advantages of each?
7. What is the main difference in the construction of a digital and an analog communication line?
8. In some cases a digital communication line is far more effective than an analog communication line. Why is this?
9. What is the meaning of "bandwidth"?
10. What is the bandwidth of a public telephone line?
11. What is the difference between leased and public lines?

12. Higher speeds have generally been used on leased voice lines than have been used on public voice lines. Why? (More than one reason.)
*13. When are public lines preferable to leased lines in a teleprocessing system?
14. What is conditioning? When is it used? How does it work?
15. What is an acoustical coupler? What are its advantages and disadvantages as compared to a data set?
16. What are escape characters? When are they used? Give an example.
17. What is the disadvantage of using parity bits for character checking?
18. Give a definition of response time.
*19. Make a list of the factors in a transmission network that affect the response time.
*20. You are designing a transmission network that will have many geographically scattered terminals and which must have a low response time to permit speedy conversational dialogue. What approaches can you take to ensure that the response time is short but that the network cost is as low as possible?
*21. When is it desirable to have buffers in terminals? List several cases. When are unbuffered terminals adequate?
*22. What is the difference between line-switching, message-switching, and packet-switching? When are message-switching systems used?
*23. Under what circumstances would you use WATS?
*24. What is a multidrop circuit? When would you use it? When would you avoid using it?
*25. On what types of system would you recommend the use of TWX?
*26. On what types of system would you not use it?
*27. What is Telpak C? When would you employ it?
28. What are the advantages of using a simple time-division multiplexer over other means of lowering network cost?
*29. What mechanisms can offer greater network cost savings than a simple time-division multiplexer? Why?
*30. Most mechanisms other than simple multiplexing for lowering network cost entail certain risks, such as the risk of obtaining "busy" signals. What different risks are associated with different mechanisms?
*31. What tariffs and mechanisms would you consider for the nine cases indicated in the table below?
    (a) For teletype machines.
    (b) For visual display units.

Number of different places called or from which calls originate:	Number of calls per day: FEW	MANY	VERY MANY
FEW			
MANY			
VERY MANY			

*32. What are the functions required at the computer center for transmission control?

*33. What functions are required in transmission control software:
(a) for simple point-to-point transmission?
(b) on a system with a few terminals?
(c) on a system with very many terminals?

*34. What types of design calculations are necessary when planning teleprocessing systems?

GLOSSARY

There is little point in redefining the wheel, and where useful the definitions in this glossary have been taken from other recognized glossaries.

A suffix "2" after a definition below indicates that it is the CCITT definition, published in *List of Definitions of Essential Telecommunication Terms,* International Telecommunication Union, Geneva.

A suffix "1" after a definition below indicates that the definition is taken from the *Data Communications Glossary,* International Business Machines Corporation, Poughkeepsie 1967 (Manual number C20–1666).

Address. A coded representation of the destination of data, or of their originating terminal. Multiple terminals on one communication line, for example, must have unique addresses. Telegraph messages reaching a switching center carry an address before their text to indicate the destination of the message.

Alphabet (telegraph or data). A table of correspondence between an agreed set of characters and the signals which represent them. (2).

Alternate routing. An alternative communications path used if the normal one is not available. These may be one or more possible alternative paths.

Amplitude modulation. One of three ways of modifying a sine wave signal in order to make it "carry" information. The sine wave, or "carrier," has its amplitude modified in accordance with the information to be transmitted.

Analog data. Data in the form of *continuously variable* physical quantities. (Compare with **Digital data.**) (1).

Analog transmission. Transmission of a continuously variable signal as opposed to a discretely variable signal. Physical quantities such as temperature are continuously variable and so are described as "analog." Data characters, on the other hand, are coded in discrete separate pulses or signal levels, and are referred to as "digital." The normal way of transmitting a telephone, or voice, signal has been analog; but now digital encoding (using PCM) is coming into use over trunks.

Application program. The working programs in a system may be classed as *application programs* and *supervisory programs*. The application programs are the main data-processing programs. They contain no input-output coding except in the form of macroinstructions that transfer control to the supervisory programs. They are usually unique to one type of application, whereas the supervisory programs could be used for a variety of different application types. A number of different terms are used for these two classes of program.

ARQ (Automatic Request for Repetition). A system employing an error-detecting code and so conceived that any false signal initiates a repetition of the transmission of the character incorrectly received. (2).

ASCII (American Standard Code for Information Interchange). Usually pronounced "ask′-ee." An eight-level code for data transfer adopted by the American Standards Association to achieve compatibility between data devices. (1).

Asynchronous transmission. Transmission in which each information character, or sometimes each word or small block, is individually synchronized, usually by the use of start and stop elements. The gap between each character (or word) is not of a necessarily fixed length. (Compare with **Synchronous transmission.**) Asynchronous transmission is also called *start-stop transmission.*

Attended operation. In data set applications, individuals are required at both stations to establish the connection and transfer the data sets from talk (voice) mode to data mode. (Compare **Unattended operation.**) (1).

Attenuation. Decrease in magnitude of current, voltage, or power of a signal in transmission between points. May be expressed in decibels. (1).

Attenuation equalizer. (See **Equalizer.**)

Audio frequencies. Frequencies that can be heard by the human ear (usually 30 to 20,000 cycles per second). (1).

Automatic calling unit (ACU). A dialing device supplied by the communications common carrier, which permits a business machine to automatically dial calls over the communication networks. (1).

Automatic dialing unit (ADU). A device capable of automatically generating dialing digits. (Compare with **Automatic calling unit.**) (1).

Bandwidth. The range of frequencies available for signaling. The differences expressed in cycles per second (hertz) between the highest and lowest frequencies of a band.

Baseband signaling. Transmission of a signal at its original frequencies, i.e., a signal not changed by modulation.

Baud. Unit of signaling speed. The speed in bauds is the number of discrete conditions or signal events per second. (This is applied only to the actual signals on a communication line.) If each signal event represents only one bit condition, baud is the same as bits per second. When each signal event represents other than one bit (e.g., see **Dibit**), baud does not equal bits per second. (1).

Baudot code. A code for the transmission of data in which five equal-length bits represent one character. This code is used in most DC teletypewriter machines where 1 start element and 1.42 stop elements are added. (See page 109.) (1).

Bel. Ten decibels, q.v.

BEX. Broadband exchange, q.v.

Bias distortion. In teletypewriter applications, the uniform shifting of the beginning of all marking pulses from their proper positions in relation to the beginning of the start pulse. (1).

Bias distortion, asymmetrical distortion. Distortion affecting a two-condition (or binary) modulation (or restitution) in which all the significant conditions have longer or shorter durations than the corresponding theoretical durations. (2).

Bit. Contraction of "binary digit," the smallest unit of information in a binary system. A bit represents the choice between a mark or space (one or zero) condition.

Bit rate. The speed at which bits are transmitted, usually expressed in bits per second. (Compare with **Baud.**)

Broadband. Communication channel having a bandwidth greater than a voice-grade channel, and therefore capable of higher-speed data transmission. (1).

Broadband exchange (BEX). Public switched communication system of Western Union, featuring various bandwidth FDX connections. (1).

Buffer. A storage device used to compensate for a difference in rate of data flow, or time of occurrence of events, when transmitting data from one device to another. (1).

Cable. Assembly of one or more conductors within an enveloping protective sheath, so constructed as to permit the use of conductors separately or in groups. (1).

Carrier. A continuous frequency capable of being modulated, or impressed with a second (information carrying) signal. (1).

Carrier, communications common. A company which furnishes communications services to the general public, and which is regulated by appropriate local, state, or federal agencies. The term strictly includes truckers and movers, bus lines, and airlines, but is usually used to refer to telecommunication companies.

Carrier system. A means of obtaining a number of channels over a single path by modulating each channel on a different carrier frequency and demodulating at the receiving point to restore the signals to their original form.

Carrier telegraphy, carrier current telegraphy. A method of transmission in which the signals from a telegraph transmitter modulate an alternating current. (2).

Central office. The place where communications common carriers terminate customer lines and locate the switching equipment which interconnects those lines. (Also referred to as an *exchange, end office,* and *local central office.*)

Chad. The material removed when forming a hole or notch in a storage medium such as punched tape or punched cards.

Chadless tape. Perforated tape with the chad partially attached, to facilitate interpretive printing on the tape.

Channel. 1. (CCITT and ASA standard) A means of one-way transmission. (Compare with **Circuit.**)
2. (Tariff and common usage) As used in the tariffs, a path for electrical transmission between two or more points without common-carrier-provided terminal equipment. Also called *circuit, line, link, path,* or *facility.* (1).

Channel, analog. A channel on which the information transmitted can take any value between the limits defined by the channel. Most voice channels are analog channels.

Channel, voice-grade. A channel suitable for transmission of speech, digital or analog data, or facsimile, generally with a frequency range of about 300 to 3400 cycles per second.

12-channel group (of carrier current system). The assembly of 12 telephone channels, in a carrier system, occupying adjacent bands in the spectrum, for the purpose of simultaneous modulation or demodulation. (2).

Character. Letter, figure, number, punctuation or other sign contained in a message. Besides such characters, there may be characters for special symbols and some control functions. (1).

Characteristic distortion. Distortion caused by transients which, as a result of the modulation, are present in the transmission channel and depend on its transmission qualities.

Circuit. A means of both-way communication between two points, comprising associated "go" and "return" channels. (1).

Circuit, four-wire. A communication path in which four wires (two for each direction of transmission) are presented to the station equipment. (1).

Circuit, two-wire. A metallic circuit formed by two conductors insulated from each other. It is possible to use the two conductors as either a one-way transmission path, a half-duplex path, or a duplex path. (1).

Common carrier. (*See* **Carrier, communications common.**)

Compandor. A compandor is a combination of a compressor at one point in a communication path for reducing the volume *range* of signals, followed by an expandor at another point for restoring the original volume range. Usually its purpose is to improve the ratio of the signal to the interference entering in the path between the compressor and expandor. (2).

Compressor. Electronic device which compresses the volume range of a signal, used in a compandor (q.v.). An "expandor" restores the original volume range after transmission.

Conditioning. The addition of equipment to a leased voice-grade channel to provide minimum values of line characteristics required for data transmission. (1).

Contention. This is a method of line control in which the terminals request to transmit. If the channel in question is free, transmission goes ahead; if it is not free, the terminal will have to wait until it becomes free. The queue of contention requests may be built up by the computer, and this can either be in a prearranged sequence or in the sequence in which the requests are made.

Control character. A character whose occurrence in a particular context initiates, modifies, or stops a control operation—e.g., a character to control carriage return. (1).

Control mode. The state that all terminals on a line must be in to allow line control actions, or terminal selection to occur. When all terminals on a line are in the control mode, characters on the line are viewed as control characters performing line discipline, that is, polling or addressing. (1).

Cross-bar switch. A switch having a plurality of vertical paths, a plurality of horizontal paths, and electromagnetically operated mechanical means for interconnecting any one of the vertical paths with any of the horizontal paths. See page 326. (2).

Cross-bar system. A type of line-switching system which uses cross-bar switches.

Cross talk. The unwanted transfer of energy from one circuit, called the *disturbing* circuit, to another circuit, called the *disturbed* circuit. (2).

Cross talk, far-end. Cross talk which travels along the disturbed circuit in the same direction as the signals in that circuit. To determine the far-end cross talk between two pairs, 1 and 2, signals are transmitted on pair 1 at station A, and the level of cross talk is measured on pair 2 at station B. (1).

Cross talk, near-end. Cross talk which is propagated in a disturbed channel in the direction opposite to the direction of propagation of the current in the disturbing channel. Ordinarily, the terminal of the disturbed channel at which the near-end cross talk is present is near or coincides with the energized terminal of the disturbing channel. (1).

Dataphone. Both a service mark and a trademark of AT & T and the Bell System. As a service mark it indicates the transmission of data over the telephone network. As a trademark it identifies the communications equipment furnished by the Bell System for data communications services. (1).

Data set. A device which performs the modulation/demodulation and control functions necessary to provide compatibility between business machines and communications facilities. (*See also* **Line adapter, Modem,** *and* **Subset.**) (1).

Data-signaling rate. It is given by $\sum_{i=1}^{m} \frac{1}{T_i} \log_2 n_i$, where m is the number of parallel channels, T is the minimum interval for the ith channel, expressed in seconds,

n is the number of significant conditions of the modulation in the ith channel. Data-signaling rate is expressed in bits per second. (2).

Dataspeed. An AT & T marketing term for a family of medium-speed paper tape transmitting and receiving units. Similar equipment is also marketed by Western Union. (1).

DDD. (*See* **Direct distance dialing,** q.v.)

Decibel (db). A tenth of a bel. A unit for measuring relative strength of a signal parameter such as power, voltage, etc. The number of decibels is ten times the logarithm (base 10) of the ratio of the measured quantity to the reference level. The reference level must always be indicated, such as 1 milliwatt for power ratio. (1). See Fig. 9.5.

Delay distortion. Distortion occurring when the envelope delay of a circuit or system is not constant over the frequency range required for transmission.

Delay equalizer. A corrective network which is designed to make the phase delay or envelope delay of a circuit or system substantially constant over a desired frequency range. (*See* **Equalizer.**) (1).

Demodulation. The process of retrieving intelligence (data) from a modulated carrier wave; the reverse of modulation. (1).

Diagnostic programs. These are used to check equipment malfunctions and to pin-point faulty components. They may be used by the computer engineer or may be called in by the supervisory programs automatically.

Diagnostics, system. Rather than checking one individual component, system diagnostics utilize the whole system in a manner similar to its operational running. Programs resembling the operational programs will be used rather than systematic programs that run logical patterns. These will normally detect overall system malfunctions but will not isolate faulty components.

Diagnostics, unit. These are used on a conventional computer to detect faults in the various units. Separate unit diagnostics will check such items as arithmetic circuitry, transfer instructions, each input-output unit, and so on.

Dial pulse. A current interruption in the DC loop of a calling telephone. It is produced by the breaking and making of the dial pulse contacts of a calling telephone when a digit is dialed. The loop current is interrupted once for each unit of value of the digit. (1).

Dial-up. The use of a dial or pushbutton telephone to initiate a station-to-station telephone call.

Dibit. A group of two bits. In four-phase modulation, each possible dibit is encoded as one of four unique carrier phase shifts. The four possible states for a dibit are 00, 01, 10, 11.

Differential modulations. A type of modulation in which the choice of the significant condition for any signal element is dependent on the choice for the previous signal element. (2).

Digital data. Information represented by a code consisting of a sequence of discrete elements. (Compare with **Analog data.**) (1).

Digital signal. A discrete or discontinuous signal; one whose various states are discrete intervals apart. (Compare with **Analog transmission.**) (1).

Direct distance dialing (DDD). A telephone exchange service which enables the telephone user to call other subscribers outside his local area without operator assistance. In the United Kingdom and some other countries, this is called *Subscriber Trunk Dialing* (STD).

Disconnect signal. A signal transmitted from one end of a subscriber line or trunk to indicate at the other end that the established connection should be disconnected. (1).

Distortion. The unwanted change in waveform that occurs between two points in a transmission system. (1).

Distributing frame. A structure for terminating permanent wires of a telephone central office, private branch exchange, or private exchange, and for permitting the easy change of connections between them by means of cross-connecting wires. (1).

Double-current transmission, polar direct-current system. A form of binary telegraph transmission in which positive and negative direct currents denote the significant conditions. (2).

Drop, subscriber's. The line from a telephone cable to a subscriber's building. (1).

Duplex transmission. Simultaneous two-way independent transmission in both directions. (Compare with **Half-duplex transmission.** Also called *full-duplex transmission.*) (1).

Duplexing. The use of duplicate computers, files or circuitry, so that in the event of one component failing an alternative one can enable the system to carry on its work.

Echo. An echo is a wave which has been reflected or otherwise returned with sufficient magnitude and delay for it to be perceptible in some manner as a wave distinct from that directly transmitted.

Echo check. A method of checking data transmission accuracy whereby the received data are returned to the sending end for comparison with the original data.

Echo suppressor. A line device used to prevent energy from being reflected back (echoed) to the transmitter. It attenuates the transmission path in one direction while signals are being passed in the other direction. (1).

End distortion. End distortion of start-stop teletypewriter signals is the shifting of the end of all marking pulses from their proper positions in relation to the beginning of the start pulse.

End office. (*See* **Central office.**)

Equalization. Compensation for the attenuation (signal loss) increase with frequency. Its purpose is to produce a flat frequency response while the temperature remains constant. (1).

Equalizer. Any combination (usually adjustable) of coils, capacitors, and/or resistors inserted in transmission line or amplifier circuit to improve its frequency response. (1).

Equivalent four-wire system. A transmission system using frequency division to obtain full-duplex operation over only one pair of wires. (1).

Error-correcting telegraph code. An error-detecting code incorporating sufficient additional signaling elements to enable the nature of some or all of the errors to be indicated and corrected entirely at the receiving end.

Error-detecting and feedback system, decision feedback system, request repeat system, ARQ system. A system employing an error-detecting code and so arranged that a signal detected as being in error automatically initiates a request for retransmission of the signal detected as being in error. (2).

Error-detecting telegraph code. A telegraph code in which each telegraph signal conforms to specific rules of construction, so that departures from this construction in the received signals can be automatically detected. Such codes necessarily require more signaling elements than are required to convey the basic information.

ESS. (Electronic Switching System). Bell System term for computerized telephone exchange. ESS 1 is a central office. ESS 101 gives private branch exchange (PBX) switching controlled from the local central office. (*See* Chapter 19.)

Even parity check (odd parity check). This is a check which tests whether the number of digits in a group of binary digits is even (even parity check) or odd (odd parity check). (2).

Exchange. A unit established by a communications common carrier for the administration of communication service in a specified area which usually embraces a city, town, or village and its environs. It consists of one or more central offices together with the associated equipment used in furnishing communication service. (This term is often used as a synonym for "central office," q.v.)

Exchange, classes of. Class 1 (*see* **Regional center**); class 2 (*see* **Sectional center**); class 3 (*see* **Primary center**); class 4 (*see* **Toll center**); class 5 (*see* **End office**).

Exchange, private automatic (PAX). A dial telephone exchange that provides private telephone service to an organization and that does *not* allow calls to be transmitted to or from the public telephone network.

Exchange, private automatic branch (PABX). A private automatic telephone exchange that provides for the transmission of calls to and from the public telephone network.

Exchange, private branch (PBX). A manual exchange connected to the public telephone network on the user's premises and operated by an attendant supplied by the user. PBX is today commonly used to refer also to an automatic exchange.

Exchange, trunk. An exchange devoted primarily to interconnecting trunks.

Exchange service. A service permitting interconnection of any two customers' stations through the use of the exchange system.

Expandor. A transducer which for a given amplitude range or input voltages produces a larger range of output voltages. One important type of expandor employs the information from the envelope of speech signals to expand their volume range. (Compare **Compandor.**) (1).

Facsimile (FAX). A system for the transmission of images. The image is scanned at the transmitter, reconstructed at the receiving station, and duplicated on some form of paper. (1).

Fail softly. When a piece of equipment fails, the programs let the system fall back to a degraded mode of operation rather than let it fail catastrophically and give no response to its users.

Fall-back, double. Fall back in which two separate equipment failures have to be contended with.

Fall-back procedures. When the equipment develops a fault the programs operate in such a way as to circumvent this fault. This may or may not give a degraded service. Procedures necessary for fall-back may include those to switch over to an alternative computer or file, to change file addresses, to send output to a typewriter instead of a printer, to use different communication lines or bypass a faulty terminal, etc.

FCC. Federal Communications Commission, q.v.

FD or **FDX.** Full duplex. (*See* **Duplex.**)

FDM. Frequency-division multiplex, q.v.

Federal Communications Commission (FCC). A board of seven commissioners appointed by the President under the Communication Act of 1934, having the power to regulate all interstate and foreign electrical communication systems originating in the United States. (1).

Figures shift. A physical shift in a teletypewriter which enables the printing of numbers, symbols, upper-case characters, etc. (Compare with **Letters shift.**) (1).

Filter. A network designed to transmit currents of frequencies within one or more frequency bands and to attenuate currents of other frequencies. (2).

Foreign exchange service. A service which connects a customer's telephone to a telephone company central office normally not serving the customer's location. (Also applies to TWX service.) (1).

Fortuitous distortion. Distortion resulting from causes generally subject to random laws (accidental irregularities in the operation of the apparatus and of the moving parts, disturbances affecting the transmission channel, etc.). (2).

Four-wire circuit. A circuit using two pairs of conductors, one pair for the "go" channel and the other pair for the "return" channel. (2).

Four-wire equivalent circuit. A circuit using the same pair of conductors to give "go" and "return" channels by means of different carrier frequencies for the two channels. (2).

Four-wire terminating set. Hybrid arrangement by which four-wire circuits are terminated on a two-wire basis for interconnection with two-wire circuits.

Frequency-derived channel. Any of the channels obtained from multiplexing a channel by frequency division. (2).

Frequency-division multiplex. A multiplex system in which the available transmission frequency range is divided into narrower bands, each used for a separate channel. (2).

Frequency modulation. One of three ways of modifying a sine wave signal to make it "carry" information. The sine wave or "carrier" has its frequency modified in accordance with the information to be transmitted. The frequency function of the modulated wave may be continuous or discontinuous. In the latter case, two or more particular frequencies may correspond each to one significant condition.

Frequency-shift signaling, frequency-shift keying (FSK). Frequency modulation method in which the frequency is made to vary at the significant instants. 1. By smooth transitions: the modulated wave and the change in frequency are continuous at the significant instants. 2. By abrupt transitions: the modulated wave is continuous but the frequency is discontinuous at the significant instants. (2).

FSK. Frequency-shift keying, q.v.

FTS. Federal Telecommunications System.

Full-duplex (FD or FDX) **transmission.** (*See* **Duplex transmission.**)

Half-duplex (HD or HDX) **circuit.**

1. CCITT definition: A circuit designed for duplex operation, but which, on account of the nature of the terminal equipments, can be operated alternately only.

2. Definition in common usage (the normal meaning in computer literature): A circuit designed for transmission in either direction but not both directions simultaneously.

Handshaking. Exchange of predetermined signals for purposes of control when a connection is established between two data sets.

Harmonic distortion. The resultant presence of harmonic frequencies (due to nonlinear characteristics of a transmission line) in the response when a sinusoidal stimulus is applied. (1).

HD or **HDX.** Half duplex. (*See* **Half-duplex circuit.**)

Hertz (Hz). A measure of frequency or bandwidth. The same as cycles per second.

Home loop. An operation involving only those input and output units associated with the local terminal. (1).

In-house. *See* **In-plant system.**

In-plant system. A system whose parts, including remote terminals, are all situated in one building or localized area. The term is also used for communication systems spanning several buildings and sometimes covering a large distance, but in which no common carrier facilities are used.

International Telecommunication Union (ITU). The telecommunications agency of the United Nations, established to provide standardized communications procedures and practices including frequency allocation and radio regulations on a world-wide basis.

Interoffice trunk. A direct trunk between local central offices.

Intertoll trunk. A trunk between toll offices in different telephone exchanges. (1).

ITU. International Telecommunication Union, q.v.

Keyboard perforator. A perforator provided with a bank of keys, the manual depression of any one of which will cause the code of the corresponding character or function to be punched in a tape. (2).

Keyboard send/receive. A combination teletypewriter transmitter and receiver with transmission capability from keyboard only.

KSR. Keyboard send/receive, q.v.

Leased facility. A facility reserved for sole use of a single leasing customer. (*See also* **private line.**) (1).

Letters shift. A physical shift in a teletypewriter which enables the printing of alphabetic characters. Also, the name of the character which causes this shift. (*Compare* with **Figures shift.**) (1).

Line switching. Switching in which a circuit path is set up between the incoming and outgoing lines. Contrast with message switching (q.v.) in which no such physical path is established.

Link communication. The physical means of connecting one location to another for the purpose of transmitting and receiving information. (1).

Loading. Adding inductance (load coils) to a transmission line to minimize amplitude distortion. (1).

Local exchange, local central office. An exchange in which subscribers' lines terminate. (Also referred to as *end office.*)

Local line, local loop. A channel connecting the subscriber's equipment to the line terminating equipment in the central office exchange. Usually metallic circuit (either two-wire or four-wire). (1).

Longitudinal redundancy check (LRC). A system of error control based on the formation of a block check following preset rules. The check formation rule is applied in the same manner to each character. In a simple case, the LRC is created by forming a parity check on each bit position of all the characters in the block (e.g., the first bit of the LRC character creates odd parity among the one-bit positions of the characters in the block).

Loop checking, message feedback, information feedback. A method of checking the accuracy of transmission of data in which the received data are returned to the sending end for comparison with the original data, which are stored there for this purpose. (2).

LRC. Longitudinal redundancy check.

LTRS. Letters shift, q.v. (*See* **Letters shift.**)

Mark. Presence of signal. In telegraph communications a mark represents the closed condition or current flowing. A mark impulse is equivalent to a binary 1. (*See* page 394.)

Mark-hold. The normal no-traffic line condition whereby a steady mark is transmitted.

Mark-to-space transition. The transition, or switching from a marking impulse to a spacing impulse.

Mark-hold. The normal no-traffic line condition whereby a steady mark is transmitted. This may be a customer-selectable option. (Compare with **Space-hold.**) (1).

Master station. A unit having control of all other terminals on a multipoint circuit for purposes of polling and/or selection. (1).

Mean time to failure. The average length of time for which the system, or a component of the system, works without fault.

Mean time to repair. When the system, or a component of the system, develops a fault, this is the average time taken to correct the fault.

Message reference block. When more than one message in the system is being processed in parallel, an area of storage is allocated to each message and remains uniquely associated with that message for the duration of its stay in the computer. This is called the *message reference block* in this book. It will normally contain the message and data associated with it that are required for its processing. In most systems, it contains an area of working storage uniquely reserved for that message.

Message switching. The technique of receiving a message, storing it until the proper outgoing line is available, and then retransmitting. No direct connection between the incoming and outgoing lines is set up as in line switching (q.v.).

Microwave. Any electromagnetic wave in the radio-frequency spectrum above 890 megacycles per second. (1).

Modem. A contraction of "modulator-demodulator." The term may be used when the modulator and the demodulator are associated in the same signal-conversion equipment. (*See* **Modulation** *and* **Data set.**) (1).

Modulation. The process by which some characteristic of one wave is varied in accordance with another wave or signal. This technique is used in data sets and moderns to make business machine signals compatible with communications facilities. (1).

Modulation with a fixed reference. A type of modulation in which the choice of the significant condition for any signal element is based on a fixed reference. (2).

Multidrop line. Line or circuit interconnecting several stations. (Also called *multipoint line.*) (1).

Multiplex, multichannel. Use of a common channel in order to make two or more channels, either by splitting of the frequency band transmitted by the common channel into narrower bands, each of which is used to constitute a distinct channel (frequency-division multiplex), or by allotting this common channel in turn, to constitute different intermittent channels (time-division multiplex). (2).

Multiplexing. The division of a transmission facility into two or more channels either by splitting the frequency band transmitted by the channel into narrower bands, each of which is used to constitute a distinct channel (frequency-division multiplex), or by allotting this common channel to several different information channels, one at a time (time-division multiplexing). (2).

Multiplexor. A device which uses several communication channels at the same time, and transmits and receives messages and controls the communication lines. This device itself may or may not be a stored-program computer.

Multipoint line. (*See* **Multidrop line.**)

Neutral transmission. Method of transmitting teletypewriter signals, whereby a mark is represented by current on the line and a space is represented by the absence of current. By extension to tone signaling, neutral transmission is a method of signaling employing two signaling states, one of the states representing both a space condition and also the absence of any signaling. (Also called *unipolar*. Compare with **Polar transmission.**) (1).

Noise. Random electrical signals, introduced by circuit components or natural disturbances, which tend to degrade the performance of a communications channel. (1).

Off hook. Activated (in regard to a telephone set). By extension, a data set automatically answering on a public switched system is said to go "off hook." (Compare with **On hook.**) (1).

Off line. Not in the line loop. In telegraph usage, paper tapes frequently are punched "off line" and then transmitted using a paper tape transmitter.

On hook. Deactivated (in regard to a telephone set). A telephone not in use is "on hook." (1).

On line. Directly in the line loop. In telegraph usage, transmitting directly onto the line rather than, for example, perforating a tape for later transmission. (*See also* **On-line computer system.**)

On-line computer system. An on-line system may be defined as one in which the input data enter the computer directly from their point of origin and/or output data are transmitted directly to where they are used. The intermediate stages such as punching data into cards or paper tape, writing magnetic tape, or off-line printing, are largely avoided.

Open wire. A conductor separately supported above the surface of the ground—i.e., supported on insulators.

Open-wire line. A pole line whose conductors are principally in the form of open wire.

PABX. Private automatic branch exchange. (*See* **Exchange, private automatic branch.**)

Parallel transmission. Simultaneous transmission of the bits making up a character or byte, either over separate channels or on different carrier frequencies on the channel. (1). The simultaneous transmission of a certain number of signal elements constituting the same telegraph or data signal. For example, use of a code according to which each signal is characterized by a combination of 3 out of 12 frequencies simultaneously transmitted over the channel. (2).

Parity check. Addition of noninformation bits to data, making the number of ones in a grouping of bits either always even or always odd. This permits detection of bit groupings that contain single errors. It may be applied to characters, blocks, or any convenient bit grouping. (1).

Parity check, horizontal. A parity check applied to the group of certain bits from every character in a block. (*See also* **Longitudinal redundancy check.**)

Parity check, vertical. A parity check applied to the group which is all bits in one character. (Also called *vertical redundancy check.*) (1).

PAX. Private automatic exchange. (*See* **Exchange, private automatic.**)

PBX. Private branch exchange. (*See* **Exchange, private branch.**)

PCM. (*See* **Pulse code modulation.**)

PDM. (*See* **Pulse duration modulation.**)

Perforator. An instrument for the manual preparation of a perforated tape, in which telegraph signals are represented by holes punched in accordance with a predetermined code. Paper tape is prepared off line with this. (Compare with **Reperforator.**) (2).

Phantom telegraph circuit. Telegraph circuit superimposed on two physical circuits reserved for telephony. (2).

Phase distortion. (*See* **Distortion, delay.**)

Phase equalizer, delay equalizer. A delay equalizer is a corrective network which is designed to make the phase delay or envelope delay of a circuit or system substantially constant over a desired frequency range. (2).

Phase-inversion modulation. A method of phase modulation in which the two significant conditions differ in phase by π radians. (2).

Phase modulation. One of three ways of modifying a sine wave signal to make it "carry" information. The sine wave or "carrier," has its phase changed in accordance with the information to be transmitted.

Pilot model. This is a model of the system used for program testing purposes which is less complex than the complete model, e.g., the files used on a pilot model may contain a much smaller number of records than the operational files; there may be few lines and fewer terminals per line.

Polar transmission. A method for transmitting teletypewriter signals, whereby the marking signal is represented by direct current flowing in one direction and the spacing signal is represented by an equal current flowing in the opposite direction. By extension to tone signaling, polar transmission is a method of transmission employing three distinct states, two to represent a mark and a space and one to represent the absence of a signal. (Also called *bipolar*. Compare with **Neutral transmission.**)

Polling. This is a means of controlling communication lines. The communication control device will send signals to a terminal saying, "Terminal A. Have you anything to send?" if not, "Terminal B. Have you anything to send?" and so on. Polling is an alternative to contention. It makes sure that no terminal is kept waiting for a long time.

Polling list. The polling signal will usually be sent under program control. The program will have in core a list for each channel which tells the sequence in which the terminals are to be polled.

PPM. (*See* **Pulse position modulation.**)

Primary center. A control center connecting toll centers; a class 3 office. It can also serve as a toll center for its local end offices.

Private automatic branch exchange. (*See* **Exchange, private automatic branch.**)

Private automatic exchange. (*See* **Exchange, private automatic.**)

Private branch exchange (PBX). A telephone exchange serving an individual organization and having connections to a public telephone exchange. (2).

Private line. Denotes the channel and channel equipment furnished to a customer as a unit for his exclusive use, without interexchange switching arrangements. (1).

Processing, batch. A method of computer operation in which a number of similar input items are accumulated and grouped for processing.

Processing, in line. The processing of transactions as they occur, with no preliminary editing or sorting of them before they enter the system. (1).

Propagation delay. The time necessary for a signal to travel from one point on a circuit to another.

Public. Provided by a common carrier for use by many customers.

Public switched network. Any switching system that provides circuit switching to many customers. In the U.S.A. there are four such networks: Telex, TWX, telephone, and Broadband Exchange. (1).

Pulse-code modulation (PCM). Modulation of a pulse train in accordance with a code. (2).

Pulse-duration modulation (PDM) (**pulse-width modulation**) (**pulse-length modulation**). A form of pulse modulation in which the durations of pulses are varied. (2).

Pulse Modulation. Transmission of information by modulation of a pulsed, or intermittent, carrier. Pulse width, count, position, phase, and/or amplitude may be the varied characteristic.

Pulse-position modulation (PPM). A form of pulse modulation in which the positions in time of pulses are varied, without modifying their duration. (2).

Pushbutton dialing. The use of keys or pushbuttons instead of a rotary dial to generate a sequence of digits to establish a circuit connection. The signal form is usually multiple tones. (Also called *tone dialing, Touch-call, Touch-Tone.*) (1).

Real time. A real-time computer system may be defined as one that controls an environment by receiving data, processing them, and returning the results sufficiently quickly to affect the functioning of the environment at that time.

Reasonableness checks. Tests made on information reaching a real-time system or being transmitted from it to ensure that the data in question lie within a given range. It is one of the means of protecting a system from data transmission errors.

Recovery from fall-back. When the system has switched to a fall-back mode of operation and the cause of the fall-back has been removed, the system must be restored to its former condition. This is referred to as *recovery from fall-back.* The recovery process may involve updating information in the files to produce two duplicate copies of the file.

Redundancy check. An automatic or programmed check based on the systematic insertion of components or characters used especially for checking purposes. (1).

Redundant code. A code using more signal elements than necessary to represent the intrinsic information. For example, five-unit code using all the characters of International Telegraph Alphabet No. 2 is not redundant; five-unit code using only the figures in International Telegraph Alphabet No. 2 is redundant; seven-unit code using only signals made of four "space" and three "mark" elements is redundant. (2).

Reference pilot. A reference pilot is a different wave from those which transmit the telecommunication signals (telegraphy, telephony). It is used in carrier systems to facilitate the maintenance and adjustment of the carrier transmission system. (For example, automatic level regulation, synchronization of oscillators, etc.) (2).

Regenerative repeater. (*See* **Repeater, regenerative.**)

Regional center. A control center (class 1 office) connecting sectional centers of the telephone system together. Every pair of regional centers in the United States has a direct circuit group running from one center to the other. (1).

Repeater.

1. A device whereby currents received over one circuit are automatically repeated in another circuit or circuits, generally in an amplified and/or reshaped form.
2. A device used to restore signals, which have been distorted because of attenuation, to their original shape and transmission level.

Repeater, regenerative. Normally, a repeater utilized in telegraph applications. Its function is to retime and retransmit the received signal impulses restored to their original strength. These repeaters are speed- and code-sensitive and are intended for use with standard telegraph speeds and codes. (Also called *regen.*) (1).

Repeater, telegraph. A device which receives telegraph signals and automatically retransmits corresponding signals. (2).

Reperforator (receiving perforator.) A telegraph instrument in which the received signals cause the code of the corresponding characters or functions to be punched in a tape. (1).

Reperforator/transmitter (RT). A teletypewriter unit consisting of a reperforator and a tape transmitter, each independent of the other. It is used as a relaying device and is especially suitable for transforming the incoming speed to a different outgoing speed, and for temporary queuing.

Residual error rate, undetected error rate. The ratio of the number of bits, unit elements, characters or blocks incorrectly received but undetected or uncorrected by the error-control equipment, to the total number of bits, unit elements, characters or blocks sent. (2).

Response time. This is the time the system takes to react to a given input. If a message is keyed into a terminal by an operator and the reply from the computer, when it comes, is typed at the same terminal, response time may be defined as the time interval between the operator pressing the last key and the terminal typing the first letter of the reply. For different types of terminal, response time may be defined similarly. It is the interval between an event and the system's response to the event.

Ringdown. A method of signaling subscribers and operators using either a 20-cycle AC signal, a 135-cycle AC signal, or a 100-cycle signal interrupted 20 times per second. (1).

Routing. The assignment of the communications path by which a message or telephone call will reach its destination. (1).

Routing, alternate. Assignment of a secondary communications path to a destination when the primary path is unavailable. (1).

Routing indicator. An address, or group of characters, in the heading of a message defining the final circuit or terminal to which the message has to be delivered. (1).

RT. Reperforator/transmitter, q.v.

Saturation testing. Program testing with a large bulk of messages intended to bring to light those errors which will only occur very infrequently and which may be triggered by rare coincidences such as two different messages arriving at the same time.

Sectional center. A control center connecting primary centers; a class 2 office. (1).

Seek. A mechanical movement involved in locating a record in a random-access file. This may, for example, be the movement of an arm and head mechanism that is necessary before a read instruction can be given to read data in a certain location on the file.

Selection. Addressing a terminal and/or a component on a selective calling circuit. (1).

Selective calling. The ability of the transmitting station to specify which of several stations on the same line is to receive a message. (1).

Self-checking numbers. Numbers which contain redundant information so that an error in them, caused, for example, by noise on a transmission line, may be detected.

Serial transmission. Used to identify a system wherein the bits of a character occur serially in time. Implies only a single transmission channel. (Also called *serial-by-bit.*) (1). Transmission at successive intervals of signal elements constituting the same telegraph or data signal. For example, transmission of signal elements by a standard teleprinter, in accordance with International Telegraph Alphabet No. 2; telegraph transmission by a time-divided channel. (2).

Sideband. The frequency band on either the upper or lower side of the carrier frequency within which fall the frequencies produced by the process of modulation. (2).

Signal-to-noise ratio (S/N). Relative power of the signal to the noise in a channel. (1).

Simplex circuit.

1. CCITT definition: A circuit permitting the transmission of signals in either direction, but not in both simultaneously.
2. Definition in common usage (the normal meaning in computer literature): A circuit permitting transmission in one specific direction only.

Simplex mode. Operation of a communication channel in one direction only, with no capability for reversing. (1).

Simulation. This is a word which is sometimes confusing as it has three entirely different meanings, namely:

Simulation for design and monitoring. This is a technique whereby a model of the working system can be built in the form of a computer program. Special computer languages are available for producing this model. A complete system may be described by a succession of different models. These models can then be adjusted easily and endlessly, and the system that is being designed or monitored can be experimented with to test the effect of any proposed changes. The simulation model is a program that is run on a computer separate from the system that is being designed.

Simulation of input devices. This is a program testing aid. For various reasons it is undesirable to use actual lines and terminals for some of the program testing. Therefore, magnetic tape or other media may be used and read in by a special program which makes the data appear as if they came from actual lines and terminals. Simulation in this sense is the replacement of one set of equipment by another set of equipment and programs, so that the behavior is similar.

Simulation of supervisory programs. This is used for program testing purposes when the actual supervisory programs are not yet available. A comparatively simple program to bridge the gap is used instead. This type of simulation is the replacement of one set of programs by another set which imitates it.

Single-current transmission, (inverse) **neutral direct-current system.** A form of telegraph transmission effected by means of unidirectional currents. (2).

Space. 1. An impulse which, in a neutral circuit, causes the loop to open or causes absence of signal, while in a polar circuit it causes the loop current to flow in a direction opposite to that for a mark impulse. A space impulse is equivalent to a binary 0. 2. In some codes, a character which causes a printer to leave a character width with no printed symbol. (1).

Space-hold. The normal no-traffic line condition whereby a steady space is transmitted. (Compare with **Mark-hold.**) (1).

Space-to-mark transition. The transition, or switching, from a spacing impulse to a marking impulse. (1).

Spacing bias. *See* **Distortion, bias.**

Spectrum. 1. A continuous range of frequencies, usually wide in extent, within which waves have some specific common characteristic. 2. A graphical representation of the distribution of the amplitude (and sometimes phase) of the components of a wave as a function of frequency. A spectrum may be continuous or, on the contrary, contain only points corresponding to certain discrete values. (2).

Start element. The first element of a character in certain serial transmissions, used to permit synchronization. In Baudot teletypewriter operation, it is one space bit. (1).

Start-stop system. A system in which each group of code elements corresponding to an alphabetical signal is preceded by a start signal which serves to prepare the receiving mechanism for the reception and registration of a character, and is followed by a stop signal which serves to bring the receiving mechanism to rest in preparation for the reception of the next character. (Contrast with **Synchronous system.**) (Start-stop transmission is also referred to as *asynchronous transmission,* q.v.)

Station. One of the input or output points of a communications system—e.g., the telephone set in the telephone system or the point where the business machine interfaces the channel on a leased private line. (1).

Status maps. Tables which give the status of various programs, devices, input-output operations, or the status of the communication lines.

Step-by-step switch. A switch that moves in synchronism with a pulse device such as a rotary telephone dial. Each digit dialed causes the movement of successive selector switches to carry the connection forward until the desired line is reached. (Also called *stepper switch.* Compare with **Line switching** and **Cross-bar system.**) (1).

Step-by-step system. A type of line-switching system which uses step-by-step switches. (1).

Stop bit. (See **Stop element.**)

Stop element. The last element of a character in asynchronous serial transmissions, used to ensure recognition of the next start element. In Baudot teletypewriter operation it is 1.42 mark bits. (*See also* **Start-stop transmission.**) (1).

Store and forward. The interruption of data flow from the originating terminal to the designated receiver by storing the information enroute and forwarding it at a later time. (*See* **Message switching.**)

Stunt box. A device to 1. control the nonprinting functions of a teletypewriter terminal, such as carriage return and line feed; and 2. a device to recognize line control characters (e.g., DCC, TSC, etc.). (1).

Subscriber trunk dialing. (*See* **direct distance dialing.**)

Subscriber's line. The telephone line connecting the exchange to the subscriber's station. (2).

Subscriber's loop. (*See* **Local loop.**)

Subset. A subscriber set of equipment, such as a telephone. A modulation and demodulation device. (Also called *data set,* which is a more precise term.) (1).

Subscriber's loop. (*See* **Local loop.**)

Subvoice-grade channel. A channel of bandwidth narrower than that of voice-grade channels. Such channels are usually subchannels of a voice-grade line. (1).

Supergroup. The assembly of five 12-channel groups, occupying adjacent bands in the spectrum, for the purpose of simultaneous modulation or demodulation. (2).

Supervisory programs. Those computer programs designed to coordinate service and augment the machine components of the system, and coordinate and service application programs. They handle work scheduling, input-output operations, error actions, and other functions.

Supervisory signals. Signals used to indicate the various operating states of circuit combinations. (1).

Supervisory system. The complete set of supervisory programs used on a given system.

Support programs. The ultimate operational system consists of supervisory programs and application programs. However, a third set of programs are needed to install the system, including diagnostics, testing aids, data generator programs, terminal simulators, etc. These are referred to as *support programs.*

Suppressed carrier transmission. That method of communication in which the carrier frequency is suppressed either partially or to the maximum degree possible. One or both of the sidebands may be transmitted. (1).

Switch hook. A switch on a telephone set, associated with the structure supporting the receiver or handset. It is operated by the removal or replacement of the receiver or handset on the support. (*See also* **Off hook** *and* **On hook.**) (1).

Switching center. A location which terminates multiple circuits and is capable of interconnecting circuits or transferring traffic between circuits; may be automatic, semiautomatic, or torn-tape. (The latter is a location where operators tear off the incoming printed and punched paper tape and transfer it manually to the proper outgoing circuit.) (1).

Switching message. (*See* **Message switching.**)

Switchover. When a failure occurs in the equipment a switch may occur to an alternative component. This may be, for example, an alternative file unit, an alternative communication line or an alternative computer. The switchover process may be automatic under program control or it may be manual.

Synchronous. Having a constant time interval between successive bits, characters, or vents. The term implies that all equipment in the system is in step.

Synchronous system. A system in which the sending and receiving instruments are operating continuously at substantially the same frequency and are maintained, by means of correction, if necessary, in a desired phase relationship. (Contrast with **Start-stop system.**) (2).

Synchronous transmission. A transmission process such that between any two significant instants there is always an integral number of unit intervals. (Contrast with **Asynchronous** or **Start-stop transmission.**) (1).

Tandem office. An office that is used to interconnect the local end offices over tandem trunks in a densely settled exchange area where it is uneconomical for a telephone company to provide direct interconnection between all end offices. The tandem office completes all calls between the end offices but is not directly connected to subscribers. (1).

Tandem office, tandem central office. A central office used primarily as a switching point for traffic between other central offices. (2).

Tariff. The published rate for a specific unit of equipment, facility, or type of service provided by a communications common carrier. Also the vehicle by which the regulating agencies approve or disapprove such facilities or services. Thus the tariff becomes a contract between customer and common carrier.

TD. Transmitter-distributor, q.v.

Teleprocessing. A form of information handling in which a data-processing system utilizes communication facilities. (Originally, but no longer, an IBM trademark.) (1).

Teletype. Trademark of Teletype Corporation, usually referring to a series of different types of teleprinter equipment such as tape punches, reperforators, page printers, etc., utilized for communications systems.

Teletypewriter exchange service (TWX). An AT&T public switched teletypewriter service in which suitably arranged teletypewriter stations are provided with lines to a central office for access to other such stations throughout the U.S.A. and Canada. Both Baudot- and ASCII-coded machines are used. Business machines may also be used, with certain restrictions. (1).

Telex service. A dial-up telegraph service enabling its subscribers to communicate directly and temporarily among themselves by means of start-stop apparatus and of circuits of the public telegraph network. The service operates worldwide. Baudot equipment is used. Computers can be connected to the Telex network.

Terminal. Any device capable of sending and/or receiving information over a communication channel. The means by which data are entered into a computer system and by which the decisions of the system are communicated to the environment it affects. A wide variety of terminal devices have been built, including teleprinters, special keyboards, light displays, cathode tubes, thermocouples, pressure gauges and other instrumentation, radar units, telephones, etc.

TEX. (*See* **Telex service.**)

Tie line. A private-line communications channel of the type provided by communications common carriers for linking two or more points together.

Time-derived channel. Any of the channels obtained from multiplexing a channel by time division.

Time-division multiplex. A system in which a channel is established in connecting intermittently, generally at regular intervals and by means of an automatic distribution, its terminal equipment to a common channel. At times when these connections are not established, the section of the common channel between the distributors can be utilized in order to establish other similar channels, in turn.

Toll center. Basic toll switching entity; a central office where channels and toll message circuits terminate. While this is usually one particular central office in a city, larger cities may have several central offices where toll message circuits terminate. A class 4 office. (Also called "toll office" and "toll point.") (1).

Toll circuit (American). *See* **Trunk circuit** (British).

Toll switching trunk (American). *See* **Trunk junction** (British).

Tone dialing. (*See* **Pushbutton dialing.**)

Touch-call. Proprietary term of GT&E. (*See* **Pushbutton dialing.**)

Touch-tone. AT&T term for pushbutton dialing, q.v.

Transceiver. A terminal that can transmit and receive traffic.

Translator. A device that converts information from one system of representation into equivalent information in another system of representation. In telephone equipment, it is the device that converts dialed digits into call-routing information. (1).

Transmitter-distributor (TD). The device in a teletypewriter terminal which makes and breaks the line in timed sequence. Modern usage of the term refers to a paper tape transmitter.

Transreceiver. A terminal that can transmit and receive traffic. (1).

Trunk circuit (British), **toll circuit** (American). A circuit connecting two exchanges in different localities. *Note:* In Great Britain, a trunk circuit is approximately 15 miles long or more. A circuit connecting two exchanges less than 15 miles apart is called a *junction circuit.*

Trunk exchange (British), **toll office** (American). An exchange with the function of controlling the switching of trunk (British) [toll (American)] traffic.

Trunk group. Those trunks between two points both of which are switching centers and/or individual message distribution points, and which employ the same multiplex terminal equipment.

Trunk junction (British), **toll switching trunk** (American). A line connecting a trunk exchange to a local exchange and permitting a trunk operator to call a subscriber to establish a trunk call.

Unattended operations. The automatic features of a station's operation permit the transmission and reception of messages on an unattended basis. (1).

Vertical parity (redundancy) check. (*See* **Parity check, vertical.**)

VOGAD (Voice-Operated Gain-Adjusting Device). A device somewhat similar to a compandor and used on some radio systems; a voice-operated device which removes fluctuation from input speech and sends it out at a constant level. No restoring device is needed at the receiving end. (1).

Voice-frequency, telephone-frequency. Any frequency within that part of the audio-frequency range essential for the transmission of speech of commercial quality, i.e., 300–3400 c/s. (2).

Voice-frequency carrier telegraphy. That form of carrier telegraphy in which the carrier currents have frequencies such that the modulated currents may be transmitted over a voice-frequency telephone channel. (1).

Voice-frequency multichannel telegraphy. Telegraphy using two or more carrier currents the frequencies of which are within the voice-frequency range. Voice-frequency telegraph systems permit the transmission of up to 24 channels over a single circuit by use of frequency-division multiplexing.

Voice-grade channel. (*See* **Channel, voice-grade.**)

Voice-operated device. A device used on a telephone circuit to permit the presence of telephone currents to effect a desired control. Such a device is used in most echo suppressors. (1).

VRC. Vertical redundancy check. (*See also* **Parity check.**)

Watchdog timer. This is a timer which is set by the program. It interrupts the program after a given period of time, e.g., one second. This will prevent the system from going into an endless loop due to a program error, or becoming idle because of an equipment fault. The Watchdog timer may sound a horn or cause a computer interrupt if such a fault is detected.

WATS (Wide Area Telephone Service). A service provided by telephone companies in the United States which permits a customer by use of an access line to make calls to telephones in a specific zone in a dial basis for a flat monthly charge. Monthly charges are based on the size of the area in which the calls are placed, not on the number or length of calls. Under the WATS arrangement, the U.S. is divided into six zones to be called on a full-time or measured-time basis. (1).

Word. 1. In telegraphy, six operations or characters (five characters plus one space). ("Group" is also used in place of "word.") 2. In computing, a sequence of bits or characters treated as a unit and capable of being stored in one computer location. (1).

WPM (Words per minute). A common measure of speed in telegraph systems.

INDEX